# CONSTRUCTION SUPERVISION
# QC + HSE MANAGEMENT IN PRACTICE

# CONSTRUCTION SUPERVISION QC + HSE MANAGEMENT IN PRACTICE

Quality Control, OHS, and Environmental Performance Reference Guide

**MARK URIZAR**
FAIA, B.ARCH, PMP, MBA, MAPPSC, LEED-AP

**EL-SAYED ABDEL MONEM SAYED ABDEL HALIM**
BSC CIVIL, PMP, MAPM, PE

Copyright © 2015 by Mark Urizar; El-Sayed Abdel Monem Sayed Abdel Halim.

Library of Congress Control Number: 2015901823
ISBN: Softcover 978-1-5035-0237-6
eBook 978-1-5035-0238-3

All rights reserved. No part of this book may be reproduced or transmitted in any form or by any means, electronic or mechanical, including photocopying, recording, or by any information storage and retrieval system, without permission in writing from the copyright owner.

Any people depicted in stock imagery provided by Thinkstock are models, and such images are being used for illustrative purposes only. Certain stock imagery © Thinkstock.

Print information available on the last page.

Rev. date: 02/24/2015

To order additional copies of this book, contact:
Xlibris
1-800-455-039
www.Xlibris.com.au
Orders@Xlibris.com.au

610092

# Contents

Foreword .................................................................................................9
Abbreviations and Terms ......................................................................11

Chapter 1–Project Management in Construction ................................17

    Construction Projects.....................................................................19
    Construction Supervision ..............................................................21
        The CS's Role ..........................................................................22
        Monitoring, Recording, and Reporting ...............................29
        Certification of Construction Works .....................................30
    Project Knowledge Areas (Q-STC-CHRIPS) ...............................32
        Quality Management..............................................................34
        Scope Management ................................................................42
        Time (Schedule) Management ..............................................43
        Cost Management ..................................................................44
        Communication Management..............................................44
        Resource Management ..........................................................47
        Risk Management...................................................................49
    HSE (Health, Safety, and Environment) ......................................49
        HSE Control ............................................................................53

Chapter 2–6D (Data, Information, and Decision-Making) .................59

    Performance Monitoring and Management.................................61

Chapter 3–Phase 1: Pre-Construction—Initiating and Planning the
    Supervision Works .........................................................................63

    The Construction Supervision Management Plan......................64
    Purpose ...........................................................................................65
    Key Assumptions ............................................................................65

Key Inputs and Essential References ..................................................65
CS's Requirements at Each Phase of Construction .........................70
    Phase 1: Pre-Construction Phase ..................................................70
    Phase 2: The Construction Phase ..................................................71
    Phase 3: The Closing Phase ..........................................................75
Establish QSE Requirements ..............................................................76
CC-QSE Management Plan Submittal ..............................................81

## Chapter 4–Phase 2: The Construction Phase—Monitoring the Works ..................................................................87

Monitoring for Compliance ..................................................................88
Communicating with CC ....................................................................95
    Meetings ........................................................................................95
    Recording and Reporting ................................................................99
    Notices and Instructions ................................................................104
    Approval Notices ..........................................................................105
    Acceptance Notice ........................................................................105
    Permission-to-Use Notice ..............................................................106
CC Construction Phase Work Submittals ..........................................115
    ITP (Inspection and Test Plan) ....................................................116
    Shop Drawings ..............................................................................117
    SWMS (Safe Work Method Statement) ......................................118
Incidents, Accidents, and Events ........................................................120

## Chapter 5–Performance Management ..................................................123

1D (Data Gathering) ..........................................................................127
2D (Data Processing) ..........................................................................128
3D (Described Information) ..............................................................129
4D (Described-Information Processing) ............................................135
    Work Activity 4D Information Presentation Example ..............138
    Building Activity 4D Information Presentation Example ..........142
    Location Activity 4D Information Presentation Example ..........143
    Project Activity 4D Information Presentation Example ............144
5D (Defined Knowledge) ....................................................................145
6D (Decision-Making) ........................................................................147

Chapter 6–Phase 3: The Closing Phase of Construction ............154

    Cx (Commissioning) Works ............................................................155
    The CxA (Commissioning Agent) ................................................156
    The Cx Management Plan ..............................................................157
        Purpose of Cx Plan......................................................................157
        Scope of Cx Works .....................................................................158
        Key Inputs and Essential References .........................................159
        Role and Responsibilities............................................................160
        System and Equipment to Be Commissioned ..........................163
        The Cx Staged Process ...............................................................164
        Outputs........................................................................................172
        Cx Checklists and Notices..........................................................173
        Cx-CA (Corrective Action) Notice ...........................................175
    Defect-and-Omission Management................................................176
    CC Final Submittals .......................................................................179
        As-Built Drawings......................................................................179
        O&M (Operation and Maintenance) Manuals ........................179
    Completion and Post-Completion Activities.................................181

Index ...........................................................................................................187
About the Book and Authors ..................................................................191

Front cover: Labourers in the Kingdom of Saudi Arabia using a primitive yet innovative way to bend spiral stirrups for circular columns. An in-practice example of how difficult tasks are accomplished on-site.

*Help me find a place outside this world so I can move the Earth. And when I find this place, why should I share it with you? Use knowledge to find your own way there, as this, my friend, is how you succeed.*

Project Title (for single project use): ............................

# FOREWORD

Imagine being tasked with supervising the construction of a new city, erecting more than 1.5 million square meters of designed spaces. Added to this is the responsibility of ensuring complex technologies are appropriately embedded throughout the many complicated structures, all requiring compliance with strict quality standards. Such a project existed, named the King Khalid Medical City (KKMC), located in Dammam, Kingdom of Saudi Arabia, and this book was written in preparation for that project.

The size and scope of the project give a fair indication of the complexity involved and effort required. The KKMC project was valued in the billions of dollars, and the scope included the construction of many parts:

- main hospital building with latest medical services
- administration wing with training facilities and conference centre
- staff and visitor residential tower buildings
- community centre with mosque
- multi-storey car park linked to the main hospital building, administration wing, and residential towers.

The many parts of the construction scope and ultimately the 'city', like most other similar projects, had a fast-track programme and, thereby, a rapidly approaching completion date. Substantial upfront planning was therefore required, not only to ensure resources were in place and activities were appropriately scheduled, but also to ensure that processes were in place as to how data and information was to be managed. Once work started, it would be difficult to manage any change, considering that a continued focused effort would be required to ensure what was done and how it was done were in accordance to specification and, most importantly, to ensure errors were minimised and rework and delays avoided.

Key to monitoring and controlling construction works is the role of the construction supervisor. The construction supervisor (or CS as referred to in the book) is often an independent third party who is appointed to verify works, confirm what is done is in accordance with contract requirements, and ensure that the specified outputs are achieved. As explained in this book, construction supervision becomes most effective when it is appropriately interfaced and synced with both site activities and management oversight as this provides the means to make effective and ongoing performance decisions based on what is occurring on-site. And with effective decision-making, the work effort can be directed towards doing the right things at the right time.

The construction supervision practices and methodology outlined in this book are applicable to both the consultant and contractor as both are tasked with supervising the works undertaken. This book details how the supervision role could be planned and utilised, listing the many aspects that should be considered as construction works proceeds and as they are completed. It provides useful references with a checklist of what needs to be done along each step and phase of the construction path, from pre-construction to commissioning of completed works and post-construction or closing phase. It also outlines what 'raw' data to be collected from site activities; how it should be processed, interpreted, interrogated; and how it should be used to assess work performance, thereby making appropriate informed decisions. Construction supervision is, after all, a role that oversees construction works in order to ensure the right things are done to the required standard and, where required, also provide the means to make timely informed information that can ensure the required performance is achieved.

# Abbreviations and Terms

Below are the list of abbreviations and terms used throughout this book.

| | |
|---|---|
| 4M | Manpower, Materials, Machinery, and *Me'Awel* (an Arabic word for *subcontractor*), the four factors that are used to investigate and determine root causes of issues identified during construction works. |
| 6D or D Cycle | A six-step process that enables effective decision-making. The steps include data gathering (1D), data processing (2D), described information (3D), described-information processing (4D), defined knowledge (5D), and decision-making (6D). |
| ABCD | The contract documents, the key input documents to the construction process—namely, Agreement, Bill of Quantities, Construction Specification, and Drawings. |
| Acceptance/ Compliant | Terms used to confirm work is conforming to construction specification and in accordance with the quality requirements. |
| Accuracy | Term used to define the degree of correctness and alignment with the target value. |
| Approval | Term used for matters relating to agreed changes to contract, scope, costs, and/or schedule. |
| As-Builts | Contract drawings incorporating agreed changes that occurred during the construction process. These can also be referred to as Record Drawings or As-Constructed Drawings. |

| | |
|---|---|
| Audit | A systematic and independent examination conducted against a set of predefined criteria, with results used to determine whether compliance was achieved in relation to quality and HSE matters. Audits can also be used to determine whether arrangements were effectively implemented and whether these were in line with policy requirements and able to achieve the desired objectives. |
| Building Certifier | Third party tasked with undertaking compliance verification with statutory requirements and issuing permit to occupy at completion, also referred to as building surveyor. |
| CA | Corrective Action, the additional works or rework required to achieve compliance. This term can also be defined as the cost of poor quality. |
| CAdmin | Contract Administrator, the person whose role is to administer the construction contract. |
| CC | Construction Contractor or contractor |
| CO | Change Order, an instruction issued to change a Q-STC item. |
| Commissioning Plan | A document that details the schedule and resources to be allocated or sourced with required actions to complete the commissioning process. |
| Construction Drawings | Drawings that were prepared for and issued for construction, usually as part of the contract documents (see ABCD). |
| Continuous Improvement | Defined as a 'Plan, Do, Check, Act' (or PDCA) process as developed by W. Edward Deming. |
| CPI | Cost Performance Index (associated with earned value). |
| CS | Construction Supervisor, the person who undertakes the role of regularly monitoring site work and ensuring quality and HSE compliance (construction supervision work is usually not the role of Project Managers or the Contractor Administrator as these positions have management responsibilities). |

| | |
|---|---|
| Cx | Commissioning, the process of verifying and documenting building's energy-related systems and ensuring these were installed, calibrated, tested, and performed according to the predefined requirements specified in contract documents. |
| CxA | Commissioning Agency/Agent, person tasked with undertaking commissioning works on equipment and systems installed prior to commencing operations. |
| CxPI | Commissioning Pre-Inspection. |
| FPT | Functioning Performance Test. |
| Hazard | A source or situation with potential to cause harm, human injury, ill health, damage to property, damage to the environment, or a combination of these. |
| Hold Point | A point in the schedule where work may not proceed without authorisation by administering party (who may be CAdmin, CS, or Statutory Authority). Hold Points are usually denoted as 'H' on ITPs. |
| HR | Human Resources. |
| HSE | Health, Safety, and Environment. |
| HWPs | Hold and Witness Points. |
| IEQ | Indoor Environmental Quality. This relates to QSE aspect of post-construction, where the internal built environment can influence the health and well-being of occupants. IEQ requirements are generally specified within contract documents and achieved by the Cx process. |
| Inspection | Activity used to ensure conformance with specified or agreed requirements, including compliance with occupational health, safety, and environmental legislation (to ensure QSE compliance). |
| IO | Improvement Opportunity, part of quality management (continuous improvement). |
| ISO | International Organization for Standardization. This issues worldwide proprietary, industrial and commercial standards. |
| ITP | Inspection and Test Plan. |

| | |
|---|---|
| IPEMC | PMI-defined project processes that include Initiation, Planning, Execution, Monitoring, and Controlling, and Closing, commonly used to manage the lifecycle of projects. |
| JSA | Job Safety Analysis/Assessment. |
| Grade | Defines the characteristic of the product. |
| MSDS | Material Safety Data Sheets. |
| NC | Non-Conformance (or non-compliant). This is defined as not complying with agreed or specified criteria. The compliance criteria is generally pre-established and specified within the contract documents and by legislation or regulation. NC items are usually identified by surveillance, inspections, observation, and audits, where the minimum specified requirement has not been met and rectification is required to attain the specified quality or standard. |
| OHS | Occupation, Health, and Safety. |
| O&M | Operations and Maintenance. |
| RFI | Request for Inspection, notice issued by construction contractor as stipulated in contract. |
| Service Provider | Organisation providing appropriately skilled personnel to undertake certain roles, such as CxA, CS, CAdmin, and PM. |
| Shop Drawings | Documents prepared by the construction contractor to illustrate portions of the work in more detail than shown in the contract documents to enable specific items to be manufactured or constructed. |
| SMARTA | Acronym used to specify performance requirements, i.e. Specific, Measurable, Achievable, Relevant, Time Frame, and Agreed. |
| SPI | Schedule Performance Index (associated with earned value). |
| Surveillance | Periodic observation, investigation, or confirmation of contract activities to either identify unsafe practices or determine conformance with certain requirements. Surveillance occurs at a point in time and is not usually continuous as inspections. |

| | |
|---|---|
| SWMS | Safe Work Method Statement, work method statement combined with JSA (job safety assessment). |
| TAB | Testing, Adjusting, and Balancing (associated with commissioning of equipment and systems). |
| TOC | Table of Contents. |
| Tolerance | Specified range of limits, statistical boundaries that define the acceptable range of variation for item or for performance to be compliant. |
| Top 20 per cent | Derived from the Pareto principle, where the top 20 per cent issues and problems cause 80 per cent of errors. |
| PA | Preventative Action. |
| Permission to Use | Direction given to proceed, without transferring responsibility or liability. |
| PDCA | Deming's Cycle. Plan, Do, Check, Act (see also Continuous Improvement). |
| PPE | Personal Protective Equipment, items that should be worn to make work safe. This can include hard hats, safety vests, and shoes. |
| Precision | Consistency of output, achieving required tolerances with actions taken. |
| Prevention | Proactive approach which keeps errors out of the process by eliminating mistakes and potential defects. The cost of prevention is far less than the potential cost of rework associated with poor quality. |
| Project Knowledge Areas | PMI-defined aspects that include Quality, Scope, Time, Cost, Communications, HR, Risk, Integration, Procurement, and Stakeholder management[1], also noted as Q-STC-CHRIPS. |
| Project Management | The process used to manage a temporary endeavour, a project, from its commencement to when the required outcome is completed, such as a newly constructed building. |
| PM | Project Manager. |
| PMI | Project Management Institute, refer to www.pmi.org. |

---

[1] Items as defined by the PMI, referenced in standard titled PMBoK (Project Management Body of Knowledge).

| | |
|---|---|
| Project Processes | Processes defined in accordance with PMI (see also IPEMC). |
| QA | Quality Assurance (see below). |
| Q-STC-CHRIPS | PMI knowledge areas, which are Quality, Scope, Time, Cost, Communications, HR, Risks, Integration, Procurement, and Stakeholder management. |
| QSE or QC + HSE | Quality Control, Occupational Health and Safety (OHS), and Environment. |
| Quality | A term used to define the standard, attribute, or characteristic of something as measured against other things of a similar kind. |
| Quality Assurance | Process used to demonstrate the delivered work is in conformance with the stated requirements. |
| QC | Quality Control, activity used to ensure correctness and quality requirements are met. QC is part of the Quality Management process and, thereby, is influenced by the approach to quality and the management processes in place to ensure the required standard is achieved. |
| QM | Quality Manager. The position can also be referred to as Construction Supervisor. |
| Quality Metrics | Specifics used to measure quality and ensure quality control. |
| Value | For construction, it can be defined as quality over cost and time. For works, it can be assessed in terms of cost, time spent, experience applied, and degree of quality achieved with output produced. |
| Witness Point | A point in the schedule where a particular activity is at a state of completion or readiness for observation by the administering party. Witness Points are usually denoted as 'W' on ITPs. |
| Work | Activities as specified within the contract documents or as amended. Work must be compliant with QSE requirements and inclusive of all 4M aspects. It is performed by the construction contractor in order to achieve the required result. |
| WBS | Work Breakdown Structure. It describes project work at the activity level as work packages. |

# CHAPTER 1

# Project Management in Construction

A project is a temporary endeavour that is used to create a unique product, service, or result within a set period. It has a beginning that is established by a set commencement date and has a definitive end that is reached when all required works are completed and the outcome emerges. Project management is the process that ensures the project works occur efficiently, and it is a discipline that employs certain management 'tools' that facilitate the delivery of the project outcome. The management tools, as defined by PMI, include the process groups, which are as Initiation, Planning, Executing, Monitoring and controlling, and Closing phases (abbreviated to IPEMC), and the knowledge areas, which are the Quality, Scope, Time, Cost, Communications, HR, Risks, Integration, Procurement, and Stakeholder management aspects (abbreviated to Q-STC-CHRIPS).

Understanding both the IPEMC processes and Q-STC-CHRIPS knowledge areas is important as these provide the means to effectively manage project works. The IPEMC processes, in essence, relate to W. Edward Deming's 'Start, Plan, Do, Check, and Act' (with the addition of an 'end' process), and can be summarised into the following phases:

- phase 1: initiation, the start, where resources are mobilised to commence the project works and where the works is planned
- phase 2: executing and monitoring, the 'do-check-act' part of the works
- phase 3: closing or ending the works and, thereby, the project.

The initiation phase refers to project start-up activities, where decisions are made to undertake the works and mobilise key resources to commence the works. Once initiation activities are completed, the project moves to the planning stage, where it is determined how best to execute the works. As the plan is executed and work commences, then it is important to monitor the works and ensure these are undertaken as planned so progress remains as per schedule in accordance with pre-established baseline.

The planned outcome must also be structured to anticipate the possibility of issues and changes arising with respect to the Q-STC (quality, scope, time, and costs) aspects of the works. This requires the monitoring processes to incorporate data evaluation processes so the data gathered can be systematically assessed, understood, and appropriate decisions made. Similarly, when major changes are contemplated or required, then there must be sufficiently robust processes in place that allow the works (or parts of the works) to be replanned, altered, or adjusted along with the baseline. In such instance, a further analysis may also be necessary to fully determine whether it would be preferable to undertake the proposed works as a separate project.

As construction works near completion, the project moves into the closing phase, where the focus is shifted to commissioning and readying equipment and plant for operational use, completing outstanding works and omissions, and rectifying defects. Once the majority of these are completed and the client accepts the works, then the post-construction period commences. This is the final phase of the construction works, where the client transitions the built form into operations and where the maintenance period of the constructed commences. Completion of the project occurs when all defective works are rectified, omissions completed, and the liability period expires.

The knowledge areas, noted as Q-STC-CHRIPS, set the framework for the management processes. From these, the project constraints are the Q-STC and HR (quality, scope, time, cost, resources, and risks) and any other factor that may limit options, such as client-specific requirements. Set dates, where milestones must be achieved, and perceived risks are two examples of project constraints that require effective management. As further constraints are evaluated, it is important to also consider their possible influence and impact on the remaining knowledge areas. For instance, it is unlikely that the schedule (time) can shorten without increasing risks and causing negative impacts on cost and quality.

For the purposes of construction supervision, this book focuses on the Q-STC-CHR (quality, scope, time, cost, communication, resources and risk) aspects. The remaining aspects, which include integration, procurement, and stakeholder management, are not covered as these are regarded as not relevant to the supervision process. However, additional to these and essential to the construction management and supervision processes are the HSE (Health, Safety, and Environment) aspects and information management, which are considered integral parts of the knowledge areas within this book.

## Construction Projects

Construction projects are generally contracted out to suitably experienced companies which have the capacity and capability to undertake the required works. The required works are generally outlined in series of documentations that constitute the construction contract, also referred to as contract documents or ABCD (an abbreviation of Agreement, Bill of Quantities, Construction Specification, and Drawings). These set out the scope of works and requirements, specifying such aspects as the nature, terms, and conditions of the arrangement.

Once these are agreed and contract is executed, then the construction contract becomes legally enforceable, where each party has contractual obligations to fulfil. It is therefore essential that each party invest sufficient time to review the terms and conditions prior to contract execution and ensure that they are clear as to what the contract, scope, and corresponding conditions entail. Fully understanding these can reduce the likelihood of later disputes or disagreements and thereby also ensure parties to the contract become committed to achieving the required outcome.

Aside from the contractual obligations, rights, and responsibilities, the true aim of any construction contract is to complete the scoped works within scheduled time frame, allocated budget, and specified quality. Possibly the most important and crucial of these is achieving the specified quality as this will avoid delays and the added costs of rework and wasted materials.

Once the contract is executed, it becomes essential for the client to ensure the contract is effectively administered and performance monitored as only this will ensure contract conditions and contractual obligations are fully fulfilled. Similarly, it is important to maintain ongoing dialogue and cooperation between parties involved in executing the contract as each party has a responsibility to facilitate progress and, where required, impart knowledge and provide guidance as how best to proceed and complete the works within the contract-specific requirements.

Construction projects are unique, unlike any other project. Their uniqueness is due to the specific nature of construction works and the following attributes:

- Construction work cannot be purchased off the shelf or imported from elsewhere.
- Construction work generally occurs in open areas.
- Construction contracts have very specific terms such as a contract period, usually measured in years, and thereby also many associated time risks that can include technological, financial and political risks
- Construction contracts can be highly complex due to the following:

    - the large number of different disciplines involved, which can include architecture, civil, MEP, and building specialists
    - the large number of people brought together, many of which have never worked together before
    - the high degree of specialised works, requiring specialist disciplines to both construct and supervise the works
    - the use of uneducated labour
    - the use of different nationalities of workers and different work standards (which is commonplace in regions such as the Middle East).

- Each party to the construction contract has different and often conflicting interests

    - The owner is focused on delivering the project at minimum cost, highest quality, and least time.
    - The contractor is focused on maximising profits and not generally caring about quality.
    - The consultants' interest may vary and depend on arrangements with the client; if fees are based on percentage costs, then they may be incentivised to maximise their income by increasing project costs.

Each of these differences adds complexity and risks to the works. Each also influences how the works can be and are undertaken. Therefore, it is essential that the inherent complexities be taken into account prior to undertaking the construction works. To fully appreciate these requires sufficient and broad understanding as to how construction projects can and should be managed, underpinned by an understanding of the many required aspects of the works and how these should brought together.

## Construction Supervision

*Why do we need Construction Supervision, and why do we need to establish quality control?*

Construction supervision is both the client and CC's responsibility. However, as invariably the CC's predominant focus is on achieving completion of the works quickly at the lowest possible cost, the client must place some emphasis on ensuring the works are constructed as designed, documented, and are completed in compliance with the contract documents. To achieve this, the client has two options, which are either of the following:

- to allow the CC to self-certify, possibly also with random audits performed by a third party
- to employ an independent party to supervise the works as it proceeds and ensure it is compliant.

The self-certification option is of high risk. To reduce the inherent risks with this option, the client must ensure the CC is honest, capable, and suitably experienced to undertake the works and certification. The CC must have a proven ability to meet the required standards, which is best assessed by reviewing works previously completed. Equal consideration must also be given to the inclusion of special contractual provisions, such as the requirement for CC to employ a third party to review and confirm that the key stages and phases of works are compliant and the provision to have an extend liability period for works. Otherwise, it is likely that compliance will only be met for those items required by law to receive certificate to occupancy, in essence, achieving certification without quality control.

The option to employ an independent party to monitor the works can add some cost and complexity to the contractual arrangements, requiring the appointment of suitably qualified personnel (PM, CAdmin, CS); however, when considered fully, these additions are inconsequential to the potential risk of failure and overall project cost. Construction oversight provides the assurance that works undertaken will meet the specified standard and the eventual outcome will be as designed and documented and, thereby, achieve the designed operational life. And should any future building issues arise, then the records produced by the oversight team can be used to inform investigations as to whether failures were due to design or due to construction error, which is important in apportioning liability.

Construction supervision is a process used to monitor and review the works undertaken, capture relevant data of what has occurred, confirming (and ensuring) compliance and providing an insight as to how the CC is performing. To be effective, the construction supervision process must be proactive and follow the dictum 'Prevention is better than cure' as this will ensure a successful completion.

**The CS's Role**

*Informed inspection links construction with design.*

The CS's key task is to oversee construction work and ensure it is in accordance with the design and contract documents. Whilst overseeing the works, the CS is also required to ensure QSE aspects are compliant, record and report findings to the CAdmin and PM. To effectively fulfil this role, the CS must have the following:

- construction experience with knowledge of building science
- an understanding of the required sequence that construction works must undergo
- familiarity and experience with the statutory and contractual HSE requirements
- skills and experience in interpreting schedules, drawings, specifications, and review of shop and as-built drawings
- skill in anticipating possible problems, issues, and risks
- ability to systematically conduct inspection of works
- understanding of items requiring inspections and tests
- regular access to accurate vertical and horizontal (control) surveying information to check dimensions and accuracy of construction works
- ability to keep and maintain accurate and detailed supervision records.

The appointment of the CS should be based on ability to effectively fulfil role obligations and specific requirements for the works. It is also preferable that the CS also has prior experience in the same type of works as being undertaken. For QSE matters, it is, however, essential to have relevant past experience (and training) as this will ensure compliance is achieved and will facilitate the works' progress.

The CS's role can be established from the following checklist.

| CS's Role Checklist | ✓/✗ |
|---|---|
| **General Requirements**<br><br>General requirements for CS are to include the following:<br><br>- Be thoroughly familiar with the works and understand the requirements contained in contract documents (ABCD), as required undertake the following:<br>    - Regularly read and refer to the contract documents. Of importance are the agreement's special provisions and specifications as these are essential for the construction works (and it is unlikely that the CS would remember these in detail).
    - Clarify any issue, anomaly, and/or concerns with the CAdmin.
    - Study and carefully check the drawings, noting the following:
        - material requirements to ensure quality and quantity of material
        - discrepancies for clarification with CAdmin in order to reduce errors
        - quantity items necessary to assess progress
        - required methods, procedures, schedules, and manufacturer's requirements.
    - Collect reference material for specialised techniques to determine compliance requirements.
    - Attend meeting, report activities, and direct all communication through specified channels.<br><br>- Understand the role and requirements of supervision and note the following:
    - Inputted effort will be reflected in the quality of construction.
    - Errors or neglect can lead to non-compliant work and possibly even unsafe conditions.
    - Being fair and firm at start of construction is important as this sets the tone the remainder of the construction works. | |

| CS's Role Checklist | ✓/✗ |
|---|---|
| <ul><li>An incorrectly applied method or procedure must be corrected at the first time.</li><li>Conditions that may lead to unsatisfactory work should be anticipated whenever possible and when identified should be communicated (preferably in writing) to the CC at earliest opportunity to avoid waste of materials, labour, and strain relations.</li><li>During the works, the CS should not delay the CC unnecessarily nor interfere with the CC's methods unless it is evident that unacceptable work will result.</li><li>Cooperation with CC, such as conducting inspections promptly and thoroughly, is essential to facilitate the works process.</li></ul><ul><li>Monitor QSE aspects and record activities, include the following:<ul><li>Be present on-site regularly during the CC's daily working hours (in accordance with ITPs/HWPs and to ensure SWMS compliance).</li><li>Visit the testing laboratory to verify that appropriate test standards are being applied and ensure all test results are reported and recorded correctly.</li><li>Ensure adherence to the design specification and requirement, and ensure departures do not compromise the design performance targets.</li><li>Arrange issue of notices for non-compliances (NC, CA, PA).</li><li>Ensure all requirements as specified within contract documents are achieved.</li></ul></li></ul><ul><li>Monitor and document 4M items, noting the following:<ul><li>Manpower. Check the number of workers on-site per activity.</li><li>Materials. Check for compliance, damage, contamination, production materials are the same as those tested, source of materials, storage. Ensure also, where applicable, that material tags and labels are maintained.</li><li>Machinery. Check the number, type, and compliance with HSE requirements.</li></ul></li></ul> | |

| CS's Role Checklist | ✓/✗ |
|---|---|
| • Report results and progress, and if required, assist with performance assessment to determine the following:<br>　- progress payments with respect to the amount of work that CC has completed<br>　- whether works are on schedule (in accordance with baseline programme)<br>　- QSE rate of compliance, advising on rework, additional work required, time lost, number of incidents. | |
| **Contractual Requirements**<br><br>The CS should understand the role requirements as defined by the contract, which include the following:<br><br>• Responsibility to monitor the CC's site activities and check whether these are in accordance with the contract documents.<br>• Ensuring the CC supervises and conducts quality control of the works whilst appreciating that the CC is entitled to complete the works at the lowest possible cost as long as the contract requirements are fulfilled<br>• Applying good judgement with reference to contract documents.<br>　- acting impartially and exercising role independently<br>　- avoiding forming any noticeable habits of procedure that might then be anticipated by the workmen, inspections should be conducted at irregular intervals<br>　- not making additional demands on CC, which are not in accordance with the contract documents.<br><br>• Limits of authority.<br>　- CS is not to make decisions that have financial, contractual, time impact on the project as only client has the right to vary contract.<br>　- CS is not to provide design, planning or technical solutions as these are the responsibilities of other.<br>　- When CC issues clarification or request to deviate from contract documents, the CS is to seek direction from the CAdmin. | |

| CS's Role Checklist | ✓/✗ |
|---|---|
| • Providing advice and input to contractual matters.<br>  - CS evaluates decisions and requests made by client that have time or cost implication and advises on project impact.<br>  - If the contract documents permits a choice of methods to be used, the CS may offer suggestions if requested and should not arbitrarily demand that any given method be employed.<br><br>• Limiting communications with CC.<br>  - Never advise CC of what needs to be done.<br>  - Never instruct the CC to stop or suspend works unless absolutely essential as this will raise major contractual issues which can result in major claims by CC.<br><br>• If the CC proposes departure from contract documents that appear to be reasonable, the CS may accept the proposal for on-forwarding to the CAdmin for review and final decision. | |
| **Providing QSE Site Oversight**<br><br>The CS is required to review all QSE work aspects to ensure the following:<br><br>• Site establishment is appropriate, confirming the following:<br>  - HSE site requirements as assessed and as documented in management plans are placed.<br>  - Security of site, with control points having required registers, site induction programme, and records.<br>  - Amenities provided are suitable for the number of workers<br>  - First aid and emergency systems are in place.<br>  - Required site signage and notices are posted.<br>  - QC and HSE representatives are appointed.<br>  - Job-specific safety plan is approved and enacted, with SWMS/JSA's approved for all trades and works perceived as having safety risk or requiring work permit. | |

| CS's Role Checklist | ✓/✗ |
|---|---|
| • Hazardous substances are appropriately managed, confirming the following:<br>  - Risk assessment are completed and implemented.<br>  - Hazardous substances are identified, and site register is in place.<br>  - MSDS (material safety data sheets) for all hazardous substances are available on-site.<br>  - Training and information are provided to employees.<br>  - Records are kept.<br><br>• Dangerous goods are appropriately managed, confirming the following:<br>  - Dangerous goods are identified and management requirements in place.<br>  - Flammable and combustible liquids are identified and stored, including explosives and LPG.<br>  - Activities such as welding have SWMS approved in place prior to commencement.<br><br>• Public safety is maintained, confirming the following:<br>  - Compliance with local government requirements.<br>  - Site security.<br>  - Traffic management plan approved and implemented.<br>  - Public is protected from dust and debris.<br>  - Work on public space is properly managed.<br>  - Public liability insurance coverage in place.<br><br>• PPE (personal protection equipment) requirements are enforced, confirming the following:<br>  - Worker PPE needs are determined for each activity.<br>  - PPE signs are posted where necessary.<br>  - PPE is provided with needs regularly reviewed.<br>  - Workers understand the need for and are trained to use PPE.<br>  - PPE application is adequately monitored, with regular inspection, and PPE is replaced as necessary. | |

| CS's Role Checklist | ✓/✗ |
|---|---|
| • Site control is maintained, confirming the following:<br>　- Access routes and disposal area arrangements are implemented as approved.<br>　- Stockpiling locations are suitable and approved.<br>　- Required environmental control measures are in place.<br>　- Excavations are safe and compliant by ensuring the following:<br>　　o Proposals for deep excavations, shoring, and cofferdams are approved by suitably qualified registered professional.<br>　　o Excavation and in situ material are monitored and protected where exposed to weather and water.<br>　　o Elevation and dimensions are surveyed for compliance, with any excavation extending beyond required depth reviewed and recorded.<br>　　o Cutting or filling of slopes is minimised.<br>　　o Safety meeting held prior to commencing deep excavations, shoring, or cofferdams to review, identify hazards, and confirm safety procedure.<br>　　o Footing excavations are approved (ensuring bottom of excavation is compacted soil) prior to the placement of concrete.<br>• Site personnel certification and qualification requirements are enforced by ensuring the following:<br>　- Site register is in place.<br>　- Site personnel, such as scaffolders, dogmen, riggers, crane operators, hoist operators, forklift operators, earthmoving-equipment operators, welders, plumbers, and electricians, are certified and licensed to undertake work.<br>　- All trainees are adequately supervised by qualified assessors. | |

**Example of CS's Review of Temporary Supports**

The CC should be required to submit temporary support works proposal for review and permission to use prior to commencement. The CS should then be tasked with reviewing the submittal proposal, ensuring it includes the following:

- condition assessment report of existing structure noting support requirements, damage to be rectified, cracking, and areas where potential failure may occur
- hazard and risk assessment with appropriate mitigation measures so it is in compliance with QSE requirements
- shop drawings of proposed supports with proposed fixings to structure
- engineering certificate of temporary support.

This example highlights that there are inherent risks associated with each part of the construction works. During installation and the works, the CC should therefore be required to provide ongoing and confirmatory information of adequacy of proposal, detailing required changes or remedial action undertaken, accompanied with engineering certificate.

A suitably experienced CS with sufficient knowledge of the required works should be confident to take a firm stand on something known to be deficient or non-compliant. Firmness with immediate action is most often required to prevent non-compliances from occurring. If non-compliances are allowed to occur, the required corrective action may consume substantial resources (time, cost, and labour) and more significantly so if these are repeated or covered up for later discovery. It is therefore important for both the CS and CC to remain focused on ensuring the works are undertaken and completed as required, first time and every time.

**Monitoring, Recording, and Reporting**

The CS work record form should include the following information.

| CS Work Record Checklist | | |
|---|---|---|
| Checking dates: ......................... | | By: ......................... |
| Item Identifier: | | |
| **Supervisor Items** | **Items to Be Checked** | **Supervisor Notes** |
| **ABCD Items** | | |
| Agreement | Schedule, compliance requirements, notices, and approvals | |
| Bill of Quantities | Quantity, quality, and cost | |
| Construction Specification | Standards, codes, specialisation, coordination, tolerance, specific requirements | |

| | | |
|---|---|---|
| **CS Work Record Checklist** | | |
| Drawings | Configuration, height, level, placement, coordinates | |
| **4M Items** | | |
| Manpower | No. of persons on-site per task, management and control, CC's actual and planned resources, resource-loading histograms | |
| Materials | Quantity, size, grade, uniformity, source, supplier, tested and certified, placement | |
| Machinery | Equipment, capacity, maintenance, licence required to use, maintenance, safety | |
| Subcontractors | Coordination between trades, work permit | |
| **Q-STC Items** | | |
| ITP, Samples, Testing | | |
| **HSE Items** | | |
| Occupational Health | SWMS, PPE, induction | |
| Safety | Equipment and tagging, fire protection, barriers, work permit | |
| Environment | Site control: clean site, site traffic access and planning, noise, vibration, dust, asbestos, air quality, water quality, discharge to public sewers and natural watercourse, hazardous waste | |

## Certification of Construction Works

*In most instances, third party certification of new construction works is necessary whilst monitoring and quality control is optional. The construction supervisor ensures the latter is achieved, whilst the building surveyor ensure the built form complies with statutory codes, regulations, and its requirements.*

A building surveyor (building certifier), unlike CS, is required to undertake certification works of the project, certify the construction works, and issue an occupancy certificate at the completion of the works. During the construction phase, the building surveyor must both manage and coordinate the administrative requirements of the building act and regulations, fulfilling the following tasks:

- Review statutory approved documents (specifications and drawings).
- Ensure built works are in compliance with approved documents.
- Review required performance for fire safety and access.
- Undertake on-site inspections, and conduct performance assessments.
- Prepare on-site inspection reports that address key performance issues as necessary.
- Conduct inspections to assess suitability for building to be occupied on a staged basis or as a whole as may be required for issue of occupancy permit.

**Specialist Input**

Invariably, during the construction process, there will be certain works that will require specialist input in order for these to be correctly assessed, tested, and thereby ensure these are in compliance with contract documents. This specialist input is in addition to the CS's oversight provided and in addition to the capabilities that would be expected from a CS.

The specialist input required will depend on the type and size of the project and the available skills within the supervision team, which may necessitate the inclusion skills relating to the following:

- soils and earthwork
- concrete and asphalt
- structural steel, bolt-up
- welding works, non-destructive examination and testing
- electrical testing
- hazardous material identification and testing
- compliance and certification, including audit for testing
- plant and equipment Cx (commissioning) to provide oversight or witness of commissioning of equipment and essential services subsystems.

Allowance should therefore be made to employ all necessary specialists to support the CS in completing the works and to assist the building surveyor to certify the works as compliant. Where required, the appointed specialist should prepare all necessary inspection reports to support the certification process by confirming compliance. This may become essential in order to obtain the required permits and certification of the works.

## Project Knowledge Areas (Q-STC-CHRIPS)

The PMI's Q-STC-CHRIPS aspects define the main areas that must be considered when undertaking project work. The effective management of these ten aspects is essential to achieve project success. Key to construction works is the Q-STC (quality, scope, time, and cost) aspects as these both influence and constrain how and when works are to proceed and be undertaken. These aspects, therefore, require careful and effective management. The table below summarises the oversight responsibilities of all the knowledge aspects for construction works.

| Knowledge Area | | Phase 1 Initiation/ Planning | Phase 2 Execution | Monitoring and Verifying Works | Change Management | Phase 3 Closing |
|---|---|---|---|---|---|---|
| Q | Quality | PM | CAdmin/CS | CS, CxA | PM/CAdmin | CS, CxA |
| S | Scope | PM | CAdmin/CS | CS, CxA | PM/CAdmin | CS, CxA |
| T | Time | PM | CAdmin | CAdmin | PM/CAdmin | CAdmin |
| C | Cost | PM | CAdmin | CAdmin | PM/CAdmin | CAdmin |
| C | Communications | PM | PM | PM | PM/CAdmin | PM |
| H | Human Resources | PM | PM/CAdmin | PM/CAdmin | PM/CAdmin | PM/CAdmin |
| R | Risk | PM | PM/CAdmin | PM/CAdmin | PM/CAdmin | PM/CAdmin |
| I | Integration | PM | PM | PM | PM/CAdmin | PM |
| P | Procurement | PM | PM/CAdmin | PM/CAdmin | PM/CAdmin | PM/CAdmin |
| S | Stakeholder Management | PM | PM | PM | PM | PM |
| HSE | | PM | CAdmin/CS | CAdmin/CS | CAdmin/CS | CAdmin/CS |

Construction Supervision: QC + HSE Management in Practice

| Knowledge Area | Phase 1 | | Phase 2 | | Phase 3 |
|---|---|---|---|---|---|
| | Initiation/ Planning | Execution | Monitoring and Verifying Works | Change Management | Closing |
| Information Management | PM | PM/ CAdmin/ CS | CAdmin/CS | PM/ CAdmin/CS | PM/ CAdmin/ CS |

Figure 1.1 Construction project key areas of responsibility.

Legend:   PM: Project Manager
             CAdmin: Contract Administrator
             CS: Construction Supervisor
             CxA: Commissioning Agent

From the above figure, the CS should predominantly be focused on activities related to quality, scope, HSE, and information managements during the execution, monitoring, and closing phases. It is important that this focus not be placed or shifted to the time or cost aspects as this will invariably influence how both quality and scope are overseen.

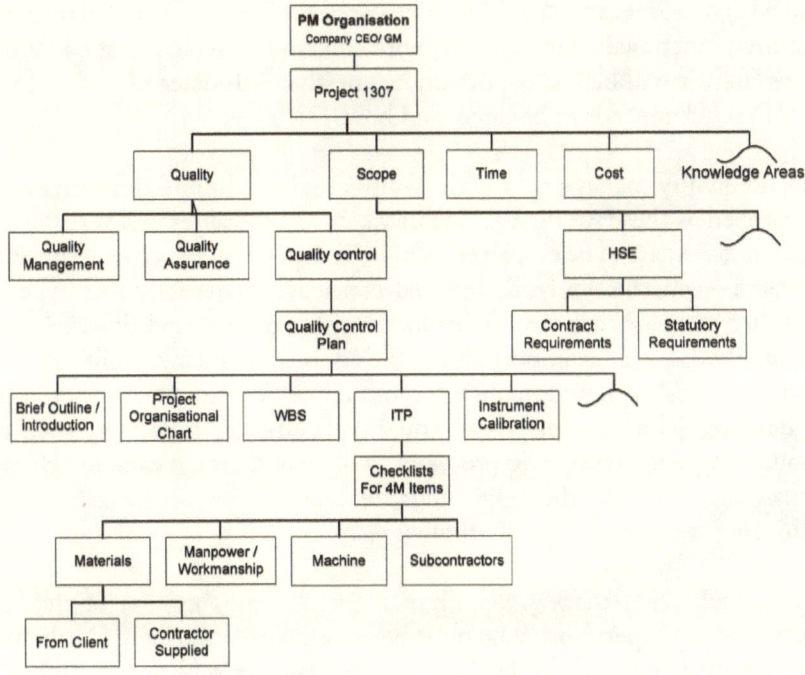

Figure 1.2 The Q-STC knowledge areas with key areas highlighted requiring CS oversight.

## Quality Management

*Quality is a constraint that must be carefully planned and managed, preferably achieved through processes rather than inspection.*

The purpose of quality management is ultimately to increase certainty with what is being produced. It reduces the amount of required rework and the risk of failure, and it provides the opportunity to improve output performance with continuous improvement.

To achieve the required quality, everyone must participate. Whilst managers, such as the PM, CAdmin, and CC (construction contractor), are responsible for product output quality (to varying degrees), everyone else involved in works, from individuals to teams, is required to perform and produce to the required (specified) standard of work. It is therefore essential from the outset to establish and convey the work-specific requirements and standards that are to be achieved as this will ensure participation in achieving these requirements.

The project-specific quality requirements should initially emerge from the design process and should later be based on decisions that balance the required functionality and amenity with constraints, such as cost and time. Inevitably, it will be the required outcome that will determine the effort required, what needs to be done, and how.

Key to quality management is the contractual conditions and systems to be applied as these set the QC (quality control) measures and verification processes that are to be employed. This will depend on the type of contract adopted—construct-only, design-and-construct, or performance type of contract—as each has a different focus with varying conditions relating to quality. Equally as important are the adopted company quality system and processes as these establish the framework for how work should be undertaken. The company QA (quality assurance) processes, such as monitoring and controlling processes, will provide the means to achieve consistency, ensuring the right things are done, improving the certainty of achieving the required standard of works.

Key to achieving the QA requirements is the appointment of the CS (construction supervisor), who must be tasked with ensuring CC achieves the specified standard. To facilitate the process and achieve the specified outcome, the CS must thoroughly understand the contract documents and know (from experience) the level of effort required, being mindful also that exceeding the stated quality requirements unnecessarily may add both cost and time to the project (also referred to as 'gold-plating' by PMI). To achieve these requirements, the CS must adopt a customised approach that is specifically tailored to suit the works.

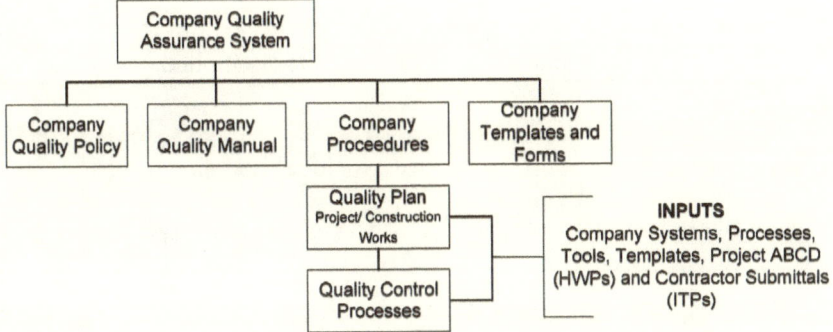

Figure 1.3 Quality management system structure.

Once the specific quality requirements of the project are established, then these must be precisely articulated in the contract documents (ABCD) as these will then set the minimum standard for the works. Once these are placed and contract is awarded, then the CC is obliged to conform to the described standards and meet at least the minimum acceptance criteria as detailed in the contract documents. And as long as compliance is achieved, the CAdmin and CS should allow the CC the freedom to choose the execution methods for the works.

**Construction Quality Plan**

The construction quality plan is an essential submittal document produced by the CC. Once received, this document should be thoroughly reviewed to ensure it covers all major aspects of the works and details how the project-specified quality requirements are to be met. This plan may not guarantee that the project will succeed, but it will ensure quality aspects are addressed in a consistent, logical, and auditable way.

The construction quality management plan should provide the following information.

| Construction Quality Management Plan Checklist | ✓/✗ |
|---|---|
| **Purpose**<br><br>Establish the purpose for the quality management plan. State the purpose, include the following:<br><br>• commitment to achieving required key quality requirements<br>• outline of quality management processes to be applied. | |
| **Key Inputs**<br><br>List key inputs and essential references, include the following:<br><br>• contract documents (ABCD), listing the quality-specific requirements, such as:<br>   - acceptance and tolerance criteria<br>   - required quality standard for work items<br>   - required approval processes for work activities (HWPs)<br>   - CC's submittals (ITPs).<br>• systems and processes to be employed, such as company QA and project plans<br>• referenced benchmark information, listing applicable requirements, such as world best practice, best practice, good practice, or common practice.[2] | |
| **Quality Management Processes**<br><br>Detail the quality management process that will be adopted, include the following:<br><br>• quality planning<br>• quality assurance<br>• quality control<br>• quality improvement. | |
| **Quality Planning**<br><br>Undertake quality planning to ensure compliance, establishing the following:<br><br>• work-specific quality assurance processes and schedules, include the following:<br>   - log of quality requirements and acceptance criteria | |

| **Construction Quality Management Plan Checklist** | ✓/✗ |
|---|---|
| <ul><li>- list of required inspections, audits, and reviews</li><li>- list of processes specified in the contract document</li><li>- required registers to keep track and record activities and processes as these are undertaken.</li></ul><br>• work-specific quality control requirements and processes<br>• change-management processes related to quality<br>• quality improvement processes. | |
| **Quality Assurance**<br><br>Adopt company or third-party system that can ensure and demonstrate compliance of work with contract requirements, ensuring the following:<br><br>• Procedures and processes to be applied include the following:<br><ul><li>- verification processes for compliance checking with specified standards and contract requirements and to validate quality processes are effective</li><li>- change-management processes to deal with changes in quality requirements and standards.</li></ul><br>• Information management procedures, include the following:<br><ul><li>- data control: procedures for recording, compiling, processing, reporting, and communicating information gained</li><li>- document control: naming and numbering convention, informative codes, and procedure for storage</li><li>- document reviews<br>   o  prior to issue<br>   o  on receipt of contractor submittals, samples, shop drawings, and as-built documentation.</li></ul><br>• Compliance review processes include the following:<br><ul><li>- activity reviews, checks at each stage against agreed methodology</li></ul> | |

---

[2] Common practice is the minimum compliance standard, complying with statutory requirements.

| Construction Quality Management Plan Checklist | ✓/✗ |
|---|---|
| <ul><li>- performance monitoring and reviews against baseline</li><li>- audits and procedures for conducting QC on work completed</li><li>- 4M resource reviews, which include appointment processes to ensure people with the right skills and relevant experience are selected and procurement processes for required equipment and materials.</li></ul><br>• System has processes that are able to identify potential problem areas and determine mitigation measures<br>• Approval processes include the following:<ul><li>- approval by authorities</li><li>- certification of works as it proceeds and at completion</li><li>- required sign-offs for works completed including for final commissioning</li><li>- required sign-offs for inspections and certifications.</li></ul><br>**Quality Control**<br><br>Set QC (quality control) measure, detailing the following:<br><br>• checklists of processes required to ensure correct implementation and to attain specified quality<br>• activities and works to be inspected<br>• processes to ensure desired quality is attained (utilise 'plan, do, check, and act' cycle), which include the following:<ul><li>- rectification process through notices and inspections, NC (non-conformance), CA (corrective action), and PA (preventative action) processes, detailing how these are actioned</li><li>- authorisation process and procedures to be followed (including for rejecting works).</li></ul><br>• testing and retesting requirements (in accordance to ITP)<br>• acceptance criteria for compliance to specific grade, accuracy, precision, and tolerance, such as for concrete, specifying concrete cube strength, concrete pullout resistance, dowel pullout resistance, coating thickness | |

| **Construction Quality Management Plan Checklist** | ✓/✗ |
|---|---|
| • evaluation process to determine effectiveness of control measures. | |
| **Quality Improvement**<br><br>Outline how continuous improvement is to be achieved, include the following:<br><br>• Identify and rectify performance issues (by issue of notices such as NC, CA)<br>• apply preventative action (PA)<br>• implement quality improvement processes (IO). | |

There are many possible benefits that can be derived from implementing a construction quality plan, including the following:

- enabling improvement, which can result in the following:
  - works completed with few defects or zero defects
  - reduction of reworking and savings in both time and cost.

- achieving effective information management by providing the following:
  - accurate data and information for more effective decision-making
  - processes that can promptly resolve emerging issues
  - processes to review compliance of works against contract documents
  - solutions such as PA and COs that are verified for buildability.

- the assurance of achieving compliance, by the following:
  - monitoring and verifying construction works until contract requirements are completed
  - commissioning and effectively planning and actioning the handover process until completed

- the avoidance of 'gold-plating' by not undertaking additional works that is not required to achieve compliance.

Possibly, the key benefit of implementing a construction quality plan is the certainty it provides that the specified quality will be met. Quality is, after all, a key client requirement and an essential item to achieve the desired outcome.

## QC (Quality Control)

*So how should construction works be monitored and controlled so specified outcomes are achieved?*

QC (quality control) is required to ensure the required work and output standards are achieved. The QC requirements can be derived from the quality management plan and compliance requirement and structured to suit both the works and company in-place systems. This provides the means for each company to the contract to use its existing and known processes, and thereby ensure the required quality is achieved. Refer to the table below for construction QC plan checklist.

| Construction QC Plan Checklist | ✓/✗ |
|---|---|
| **General Outline**<br><br>List general construction QC requirement, include the following:<br><br>• monitoring and control measures, noting the following:<br>  - established baseline requirements<br>  - processes and procedures<br>  - how and when these are to be conducted<br>  - planned audits<br>  - recording and reporting requirements.<br><br>• compliance and certification requirements. | |
| **Inputs**<br><br>List required inputs, include the following:<br><br>• contract requirements, work acceptance criteria, listing items to be inspected, HWPS<br>• company quality system, processes, and procedures (as applicable). | |

| Construction QC Plan Checklist | ✓/✗ |
|---|---|
| **QC Plan**<br><br>Ensure the QC plan includes the following:<br><br>• description of the works<br>• organisational chart, indicating roles and responsibilities and levels of authority (diagrammatic)<br>• methodology statement, indicating steps, control points, and activities<br>• ITPs, testing requirements, including instrumentation, equipment, calibration, listing tests and trials for materials and works off-site, on-site, and during commissioning phase<br>• time schedule for key activities, noting HWPS<br>• HWPS approval process and sign-off requirements<br>• vendor and subcontractor list, notification, clearances, and requirements<br>• QA/QC key personnel, experience, and qualification<br>• quality audit schedule, audit report requirements<br>• inspection schedule, including submittals requirements for pre-inspection, RFI, shop drawings<br>• reporting requirements, including the following:<br>    - inspection/fabrication status report<br>    - pre-inspection meeting reports<br>    - vendor inspection reports (for each visit conducted by contractor). | |
| **Outputs**<br><br>List CC required QC submittals, include the following:<br><br>• ITPs<br>• shop drawings<br>• RFIs and schedule. | |

Prior to starting the works, it is necessary to review work requirements, sequences of activities, and equipment and ask 'What if?' This is essential to identify problem areas, possible hazards, and also determine what, if any, preventative action should be taken (for how information is to be processed, see 5D checklist, Chapter 5). Key to these is the HSE aspects, namely the items that can potentially harm workers' health and the environment.

## Scope Management

*Scope is the work specified to complete a project.*

No two construction projects are ever the same. Therefore, substantial upfront effort is required to thoroughly understand the work and specific nature of the scope, complete with all its constraints and requirements. To do so, the total scope, 100 per cent of works should be decomposed into component parts, or work packages, limited to 80 hours duration and detailed within the schedule, with cost and quality information. This is an essential CC requirement, who should submit this information organised as WBS (work breakdown structure) for CAdmin and CS review, and for ongoing reference during the works.

*The WBS helps visualise the extent of works and thereby better understand the project with its many parts.*

Additional to each project's scope uniqueness are their specific construction contract requirements. These are the terms in which work packages and thereby the construction works are to be delivered with their Q-STC requirements (quality, scope, time, and cost). The contract documents, in essence, link the scope, the work output, with the quality and time requirements. To ascertain these terms are met, monitoring and verification of activities are necessary to ensure compliance as this will in turn determine also the cost aspect, whether a progress payment should be made or delay penalties applied.

The table below outlines the key components of contract documents.

| ABCD Summary | | |
|---|---|---|
| A | Agreement | Contract conditions, usually generic terms with project-specific items and provisions. |
| B | Bill of Quantities | Document that provides an itemised account of all work in terms of quantities, volume, numbers of materials, parts, and labour with CC's price for each item. |

---

[3] MasterFormat is a standard specification developed by the CSI (Construction Specifications Institute), suitable for commercial and institutional building projects (refer to www.csinet.org/MasterFormat).

| ABCD Summary | | |
|---|---|---|
| C | Construction Specification | Technical standard which explicitly sets the requirements for materials, services, products, and the works which must be completed as part of the contract. Specifications are generally standard and commercially available documents, such as MasterFormat[3], which were developed over many years and are precise in meaning. |
| D | Drawings | The works shown diagrammatically, detailing all dimensions, elevations, and specific requirements, such as reinforcing steel lists and bar mass. |

It is important to note that all the construction works are specified within the contract documents (ABCD). Any item not covered by these, or not included within these, should not be undertaken without written consent from the authorised party. Neither should any specified item be deleted or excluded without similar approval. Where discrepancies do occur between certain items or terms, then generally, the part that relate specifically to the works should be applied unless an alternate precedence order is specified (where priority order is specified in contract documents).

## Time (Schedule) Management

*Since a project is a temporary endeavour, it has a schedule, a known time limit and thereby, also an end date.*

When time is allocated to a project, it must be managed. Project time is managed by setting milestones within a work schedule, which also establishes the baseline for which the works must be undertaken. The project schedule is developed as the project work is decomposed into work packages and as each element is allocated time and sequenced with respect to the whole works. Once completed, this baseline can then be monitored to ensure compliance. Of vital importance is the critical path[4] for the works.

The established critical path within the schedule set the time required to complete the project. Along this path are the most critical activities, all

---

[4] Critical path activities set the longest path on the project network diagram or the shortest time required to complete the project. Any delay along this path of activities will result in an overall project delay.

requiring special care to plan, execute, monitor, and control. Potential delays that might occur from poor quality and thereby requiring rectification or rework can be avoided with effective management of these critical activities. Effective management of these can also ensure the baseline schedule is met and possibly reduce the overall project time.

As activities of work packages are established to estimate the time required to complete the work and set the schedule, so can their costs. Whilst the cost aspects are essential to the project, these do not greatly impact the supervision process. Essential to supervision is time, when activities are to occur, as this also determines when monitoring is required to ensure compliance.

**Cost Management**

The construction works' Q-STC (quality, scope, time and cost) aspects are interlinked and, thereby, difficult to manage each in isolation. Whilst cost concerns may not be part of the supervision works, these do influence quality and progress and, therefore, should be taken into consideration in all assessments undertaken. Key to these is the 4M factors (manpower, materials, machinery, and *Me'Awel*/subcontractor), which largely determines what gets done, when, how, and by whom, influencing also the quality, progress, and cost of the works.

**Communication Management**

*Construction works are reliant on effectively communicating timely relevant information.*

Communications can be formal or informal, written or verbal, and a combination of these. However, for construction works and when communicating between the different contracting parties, it is advisable that all communications be both formal (i.e. issued by authorised person) and written. It however is necessary to undertake extensive informal and verbal communications in order understand underlying causes behind emerging issues and concerns, so these can be resolved amicably and quickly. Results from informal discussions should always be confirmed in writing, preferably as minutes of meeting, listing key discussion items, decisions and instructions issued.

Typically, the most common forms of construction communications include the following:

- informal telephone or email communications—used to convey a quick message or decision
- scheduled or ad hoc meetings—held to provide a forum to inform and discuss with key stakeholders project-specific information (accomplishments, milestones, planned tasks, and issues)
- presentations—used to inform key stakeholders of major milestone events and accomplishments
- formal written communications (project reports, directions, approvals, variations, and notices)—used to inform key stakeholders of project information and decisions in accordance with established communication protocols.

Of greater importance than the form of communication is the relevance of information contained within the message, which is largely dependent on the data gathered, how it is processed, interpreted, and assessed. See the section titled '6D (Data, Information, and Decision-Making)' below. The message significant should also influence how it is to be delivered. For instance, for priority and urgent messages, these should be relayed immediately, where possible, in person, by phone, or by text and always followed up with formal notice. Invariably, the most important information will relate to contractual or Q-STC-HSE issues at hand, which must be communicated through notices and instructions as follows:

- NC (non-compliance) notice—issued to rectify QSE concerns, usually also followed with CA, PA, IO notices, requiring rectification of unacceptable works or substandard processes related to safety, site security, access, and environmental protection measures
- CO (change order) notice—issued for scope change, either to direct reduction or additional works
- time (schedule) concern notice—issued in relation to progress concerns with respect to baseline
- cost (payment or claim) notice—issued in relation to assessed value of works.

To ensure the intended message is communicated clearly and unambiguously, the SMARTA guidelines should be utilised to construct the message. The table below provides an example for communicating rectification works using the SMARTA acronym tool.

| SMARTA Example | | |
|---|---|---|
| S | Specific performance requirement. Define the expected performance target and required behaviour to achieve the target, e.g. rectify set-outs of an Olympic-size swimming pool to comply with technical specification. | |
| M | Measurable. Quantify the volume, accuracy, time, and required change, e.g. the length of the required pool is to be 50 m with tolerance not exceeding +0.03m in vertical plane. | |
| A | Achievable. Confirm that the individual or contracting firm believes that the required targeted performance can be achieved even if substantial additional effort is required, e.g. the contactor agreed to construct the Olympic-size pool within 3 months. | |
| R | Relevance. State why the effort is required and how it contributes to the overall objective, e.g. the Olympic-size pool must be constructed to international standards for it to be certified as a 'competition pool'. | |
| T | Time frame. Set the completion date for the task with required milestone reviews, e.g. weekly meetings will be held to review progress and ensure the works are completed within the required three months. | |
| A | Agreed. Ensure the individual or contracting firm agrees, e.g. the agreement can be in the form of signed minutes of the meeting with attached schedule of work showing required completion date. | |

Figure 1.4 SMARTA tool example.

Formal communications should only occur between the authorised representative from the contractor's side and client's side (CAdmin or PM) and be conducted along agreed channels. The rules for communicating formal information are the following:

- Any matter of contractual significance or substance should be formally communicated in writing.
- Verbal instructions should only be used for routine matters or for cautioning of possible failure (rectify hazardous situation).
- No instruction should be given in relations to how to perform or undertake specific work as this can lead to a contractual claim and transfer liability for that portion of construction to the administering party.
- Stop-work notices should only be issued for matter relating to high risk relating to QSE item, where there is an imminent danger of injury to person or to prevent a severe environmental incident from occurring.
- Details of communications are to be copied to all authorised persons and administrators supervising the works.

- Only in special circumstances should the CS be allowed to issue instructions or notices. In such events, the CS should immediately inform both CAdmin and PM.

Site communications should also avoid arguments; however, if these do occur, these should be immediately referred to the CAdmin for resolution. Any and all communications should be courteous and focused on the works, with criticisms reframed as caution and warning that work may not be compliant.

**Resource Management**

*For construction works, the primary focus must always remain at managing resources first followed by information management and processes.*

Invariably, it is people's actions and their productivity that ultimately determines how and when things get done and what is produced. To influence and improve productivity, the many individuals and, thereby, the team must be first coached, motivated, and then managed to ensure the required perform is achieved.

Construction work is a labour-extensive process, requiring a large number of people, many trades and professions. To reduce the overall complexity involved with the large numbers, an organisational chart is essential to diagrammatically show the many participants in terms of their place and relative rank within the structure and reporting lines with others. This chart also allows managers to easily identify, organise, coordinate the available resources to produce the required team effort that can progress the construction works.

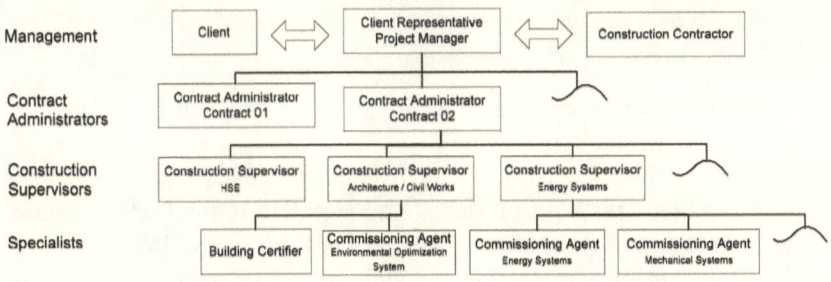

Figure 1.5 Example of a generalised organisation chart showing disciplines engaged in overseeing construction works.

The required skilled human resources are part of 4M (manpower, materials, machinery, and subcontractors), all the required resources for the works.

How the CC plans, procures, utilises, and manages these 4M resources will largely determine how the works are progressed. To facilitate the works and progress, oversight as well as guidance should be provided by the administering parties (PM, CAdmin, and CS) to the CC.

The table below lists key roles for CC (construction contractor), PM (project manager), CAdmin (Contract Administrator) and CS (construction supervisor) for each construction phase.

| Process/ Phase | Key Appointment and Roles Checklist | | | |
| --- | --- | --- | --- | --- |
| | CC | PM | CAdmin | CS |
| Phase 1: Initiation/ Planning | Mobilise 4M to site | Manage procurement process, appoint contractor | Assist in procurement process | Nil |
| Phase 2: Execution and Monitoring | Undertake construction works, Manage resources (4M), Ensure QSE compliance, and mitigate risks | Oversee construction contract, Manage HR, communications, and information | Administer construction contract | Monitor 4M and QSE to ensure compliance, record and report site progress |
| Phase 3: Closing | Undertake commissioning of equipment and plant, Complete construction works | Oversee closing process, handover, and post-construction activities; facilitate the transition to operations | Administer contract completion process and close contract | Monitor 4M, QSE, commissioning works, and defect rectifications to ensure compliance |

Figure 1.6 Roles for key appointments during construction works.

Construction projects may use many different processes and have varying delivery methods, range of products, and installation methods; however, all have in common the 4M base requirement. How the 4M aspects are employed will ultimately determine how the works are progressed, whether compliance is achieved with respect to QSE aspects and whether a timely completion will be achieved.

## Risk Management

Risk (and issue) management is generally conducted by the administering parties of the contract (CAdmin and PM). However, to ensure compliance and thereby reduce the risk of failure, work oversight is required. Oversight provides the means to ensure the required risk prevention and mitigation measures are implemented in place and the means to review works and identify emerging risks and issues. Once risks and issues are identified, only then can these be resolved, hopefully prior to these escalating to events.

The oversight role must ensure the planned is implemented and deficiencies are progressively resolved and corrected as these occur. This is the CS's role, which after all, is to monitor, verify compliance, and thereby reduce the 'risk' of failure. Ensuring compliance will inevitably increase the certainty of achieving a successful outcome.

## HSE (Health, Safety, and Environment)

As with risk management, HSE control measures are necessary to ensure both personnel safety and environmental protection. It is essential for all HSE matters to do the right things as often there are severe consequences for non-compliance. Most HSE matters are governed and enforced by law[5] (in most countries) and, furthermore, we also have moral obligation not to cause harm. Therefore, even though HSE may not be specifically included in the scope, these must be considered as essential parts of the works.

All parties associated with the construction contract should be aware of the project's HSE requirements. Everyone has some duty of care to ensure HSE compliance. There exists a chain of responsibility which starts from the worker and extends through to the CC, CS, CAdmin, PM, and the client—all who are proportionally responsible to ensure HSE compliance. Essential to HSE management is the ongoing assessment and monitoring process, which includes identifying, analysing and evaluating, responding, treating, controlling, and reporting HSE matters. This assessment process, which is similar to risk management, should be conducted by both by the client's administering parties and CC, with each party preparing their own

---

[5]  The client can be held liable for HSE compliance; however, by contract, this responsibility is assigned to the people employed, to the CC, PM, CAdmin, and CS, who must therefore ensure compliance to protect their interests.

HSE management plan and implementing measures at the pre-construction phase, prior to mobilisation, for use during the construction works.

The table below provides a checklist for establishing the HSE requirements.

| HSE Requirements Checklist | ✓/✗ |
|---|---|
| **Inputs and Key Requirements**<br><br>List key requirements from the following:<br><br>• contract documents (ABCD);<br>  - HSE performance targets, including incentives for improvement<br>  - HSE-required submittals, hazard assessment, HSE plan, and SWMS.<br><br>• legislative and statutory conditions relating to OHS and environmental protection that have been placed on the works, requiring specific action, permits, approvals, licences, certificates (note that failure to comply can result in fines and/or prosecution). | |
| **HSE Measures for Compliance**<br><br>Establish HSE measures to achieve and ensure compliance, include the following:<br><br>• Management and oversight to ensure the following:<br>  - Safe workplace with safe systems of work.<br>  - Issues, risks, incidents, and events are addressed as these occur.<br>  - Measures and processes adopted will achieve compliance and not adversely impact health or the environment.<br>  - Effective management of 4M aspects, including the following:<br>    o Manpower. Workforce is suitably experienced, trained, informed, and equipped to undertake works safely and without incident.<br>    o Materials. Appropriate measures are in place to ensure the safe delivery, storage, movement, and installation of materials.<br>    o Machinery. To ensure safe operation of site machinery. | |

| HSE Requirements Checklist | ✓/✗ |
|---|---|
| <ul><li>o Subcontractors. To ensure compliance for works contracted to others.</li></ul><ul><li>Adoption of initial HSE control measure to make site safe prior to commencing works.</li><li>Adoption of environmental safeguards to mitigate identified concerns.</li><li>Site entry requirements, ensuring that all entry points have appropriate signage displayed showing entry requirements, such as:<ul><li>- PPE (personal protective equipment)</li><li>- site induction, requiring the following:<ul><li>o all persons entering the site to be given general induction, including subcontractors, visitors, and persons carrying out non-construction type of work, covering the following:<ul><li>▪ site location information, site orientation with emergency response procedure</li><li>▪ site-specific HSE measures in place</li><li>▪ PPE (personal protective equipment) requirements, e.g. safety helmet, safety shoes, high visibility clothing</li><li>▪ any other relevant information.</li></ul></li><li>o specific inductions for work activity perceived as having high risk, given to people involved with activity</li><li>o records of each induction, where each inductee is issued with an induction card stating the inductee's full name, employer name, date of induction, and date of expiration.</li></ul></li></ul></li><li>The establishment of a site HSE committee that meet regularly, are supported by management, and are tasked with the following:<ul><li>- reviewing compliance requirements of contract, regulation, and legislation</li><li>- ensuring that all known concerns, issues, and risks are covered</li></ul></li></ul> | |

| HSE Requirements Checklist | ✓/✗ |
|---|---|
| <ul><li>ensuring that a safe and healthy worksite environment is established and maintained</li><li>encouraging employee innovation and involvement in achieving HSE compliance.</li></ul><br>• Requirement for issue-specific safety procedure as work instructions or SWMS (safe work method statement) for all tasks and activity perceived as having safety risk. Activities can include excavations, concrete works, use of tower cranes (outlining lifting capacity and precaution for lifting loads), confined-space work, cold and hot work.<br>• Requirement for CC to monitor and confirm HSE compliance, to ensure also the following:<ul><li>Non-compliances are promptly rectified.</li><li>Focus is set and remains on prevention of incidents and accidents.</li></ul><br>• Submittals of HSE records include HSE plans, SWMS, hazardous-substance register, reports, notices, and other documentation as required by contract and legislation, ensuring that these are also readily available on request.<br>• Implementation of site-specific emergency management plan to minimise the risk of injury, property, and environment damage, ensuring that plan is rehearsed on a regular basis. | |
| **Outputs**<br><br>List required outputs, including CC's submittals (hazard assessment, HSE plan, and SWMS for activities with safety risk). | |

The benefit of listing the HSE requirements upfront is the establishment of a planned approach to manage the HSE aspects and ensure compliance. Adopting a planned approach has a potential to reduce potential accidents (health related) or incidents (environmental) and, thereby, maintain worker morale, reduce sick leave and staff turnover, and possibly enhance the CC's reputation.

## HSE Control

*The aim should be set at achieving zero accident, zero near misses, and zero incident.*

We are responsible for our own personal safety and undertaking work in a safe manner; however, when at work, this responsibility also extends to others. The duty to provide a safe workplace starts with the CC and extends to overseers of the construction works, which includes the CS. The CS has the responsibility of ensuring the CC has fulfilled the obligations of providing a safe work environment by implementing all safety measures, procedures, and processes as noted in the site-specific safety plan.

To achieve a safe work environment, all site hazards and risks must first be identified and then assessed for these to be effectively managed and controlled. The assessment process must evaluate the likelihood of events occurring and develop strategies and measures that either remove or reduce hazards to safe level. Once measures are implemented, it is then important to verify implementation through monitoring. Below is the checklist for hazard assessment.

| Hazard Assessment Form Checklist | ✓/✗ |
|---|---|
| **Hazard Identifier**<br><br>The hazard assessment should include the following general information:<br><br>• hazard-item identifier number, in sequence<br>• assessor's name and date of assessment. | |
| **Details of Assessment**<br><br>Ensure sufficient detail is provided to undertake assessment, include the following:<br><br>• hazard details (when, where, who, what)<br>• hazard assessment in terms of the following:<br>   - probability (high/moderate/low)<br>   - impact (high/moderate/low)<br>   - ranking (high/moderate/low).<br>• immediate measures to be taken, such as not to commence work if unsafe or is likely to become unsafe<br>• required actions to remove or reduce hazard to safe level. | |

| Hazard Assessment Form Checklist | ✓/✗ |
|---|---|
| **Monitoring, Following Up, and Closing**<br><br>Follow up required actions during implementation of action items, include the following:<br><br>• measures to be applied and implemented<br>• monitoring and control requirements to verify hazard is brought to safe level<br>• closed-out details, rectification date, confirmed by name and signature<br>• follow-up verification details and date. | |

In determining what measures should be applied, it is important to first gauge the level of risk or hazard severity and then prioritise these in terms of severity, likelihood, and frequency. The required measures can be determined as follows:

- For high-risk items, likely to cause death or permanent injury, these hazards should be eliminated or removed from workplace.
- For medium-risk items, with possibility of causing serious injury, these hazard should be:
    - substituted, swapped with hazard of lower risk level isolated or removed from employees or employees separated from the hazard (requiring work permit to undertake activity)
    - re-engineered by physically altering the work environment to make it safer (change work practice)
    - administered by implementing SWMS, by reducing employee exposures, by job rotation and restricting access to the work area, and by enforcing PPE in accordance with SWMS.

- For low risks and hazards with potential to cause minor injury only, these should be managed accordingly, with respect to best practice.
  In assessing risk and hazard, it is important to:
- consider cultural and personnel differences, such as height, weight, language, literacy skills, age, and experience, as these may increase likelihood of occurrence
- undertake periodic additional assessments, considering the following:
    - whether any additional hazards were introduced into the workplace with measures employed or as work is progressed

- effectiveness of measures adopted by monitoring number of HSE incidents and lost time.

- implement a range of solutions, combination of measures and controls as some hazards may never be completely eliminated or replaced with safer alternatives
- prepare SWMS (safe work method statement) for all high and medium risks, with measures aimed at reducing the potential risk whilst, in contrast, low risks need only be minimised as far as possible (and need require a SWMS)
- include the administration option of issuing PPE directives for all options, but regarding these as the last choice in hazard control.

Below is checklist table for assessing construction HSE management plan.

| Construction HSE Management Plan Checklist | ✓/✗ |
|---|---|
| **General Outline**<br><br>List HSE compliance and certification requirements for the site and works, include the following:<br><br>• required HSE control measures to be implemented (to make site and work safe)<br>• legal and legislative compliance requirements (OHS and environmental protection)<br>• design and engineering solutions to be adopted to ensure safe construction<br>• management and oversight that will be adopted. | |
| **Inputs**<br><br>List required inputs, include the following:<br><br>• contract requirements, HSE performance targets<br>• hazard assessment for works<br>• organisational and project HSE system, processes, and procedures (as applicable). | |

| Construction HSE Management Plan Checklist | ✓/✗ |
|---|---|
| **HSE Management Plan**<br><br>The CC HSE plan should detail the following:<br><br>- HSE policy, with measurable objectives, signed by CEO/owner and reviewed in the last 12 months<br>- project governance structure ensuring appointment of dedicated HSE resources<br>- list of identified site hazards for planned construction works with proposed mitigation measures<br>- HSE risk assessment with responses to identified risks and hazards<br>- inspection schedule, including HSE audit schedule<br>- reporting requirements, including incident notification procedure with proposed responses<br>- general and specific HSE requirements for works, outlined as the following:<br>  - site induction requirements and procedures<br>  - site safety training<br>  - specific actions to comply with requirements and legislation, including the following:<br>    - detailed health, safety, and environmental protection management plans<br>    - implementation strategy for required measures, including staff training and monitoring to ensure compliance.<br>  - responsibilities and obligations for each person and activity, listing HSE key personnel, experience, and qualifications.<br>- 4M aspects, including the following:<br>  - manpower—requirements for site personnel (staff, labourers, employee), detailing the following:<br>    - roles and responsibilities, identifying who is responsible for what | |

| Construction HSE Management Plan Checklist | ✓/✗ |
|---|---|
| <ul><li>○ experience, qualification, and training required to undertake work</li><li>○ work information, with instructions, PPE requirements, and supervision</li><li>- materials—requirements for delivery, storage, movement, installation, and processes to ensure hazard substances are used in a safe manner with hazardous waste disposal processes</li><li>- machinery—requirements for permanent and temporary machine, maintenance, movement within site to ensure safe operation of site machinery</li><li>- subcontractors—ensure processes in place to achieve HSE compliance for works are similar to contracted parties.</li></ul><br>• environmental-specific protection measures, including the following:<br><ul><li>- compliance with the legislative requirements, noting that failure to comply can result in fines and/or prosecution</li><li>- implementation of environmental safeguards to mitigate identified concerns, risks, and delivery of work, providing measures to control (as applicable) the following:<ul><li>○ noise</li><li>○ vibration</li><li>○ dust</li><li>○ asbestos</li><li>○ site traffic access and planning</li><li>○ air quality</li><li>○ water quality</li><li>○ discharge to public sewers and natural watercourses</li><li>○ protection of endangered species</li><li>○ protection of cultural resources</li><li>○ protection of wetlands.</li></ul></li></ul><br>• site emergency response management plan, procedures, and rehearsal plans<br>• insurance coverage as required by contract. | |

| Construction HSE Management Plan Checklist | ✓/✗ |
|---|---|
| **Outputs**<br><br>List required key HSE submittals, include the following:<br><br>• hazard responsibility table detailing key appointments<br>• CC HSE management plan<br>• SWMS. | |

There are commonalities between QC (quality control), OHS (occupational health and safety), and environment control (QC + HSE or QSE). Each is key to construction works, and each requires a proactive management approach to achieve compliance. By integrating these with one management system, certain efficiencies can be gained as duplication of effort is reduced with respect to planning, implementation, monitoring, and controlling activities. It is therefore advantageous for CC to adopt an approach that is integrated and is able to effectively link quality management, safety, and environmental protection. Such an approach would ensure prompt rectification action should any non-compliance occur, including incidents or accidents, thereby also minimising any potential damage and delay.

# Chapter 2

# 6D (Data, Information, and Decision-Making)

*Knowledge is gained from many sources, from books, from teaching and gained from the experience of doing. And with new insight, the existing 'current' knowledge can be transformed into new.*

Relevant data is required to make informed decisions. For data to be relevant, it must be timely, accurate, appropriately collated, and correctly interpreted so it is transformed into meaningful information and for it to become insightful knowledge. For construction works, the CS's role is key to obtaining relevant data as the CS is, in part, tasked with recording and reporting site activities as these occur. Once data is collected, the conversion process must be sufficiently comprehensive and robust enough so data is interrogated and becomes meaningful information that is readily applicable. The figure below lists the six key steps of the 6D, or the D-cycle process.

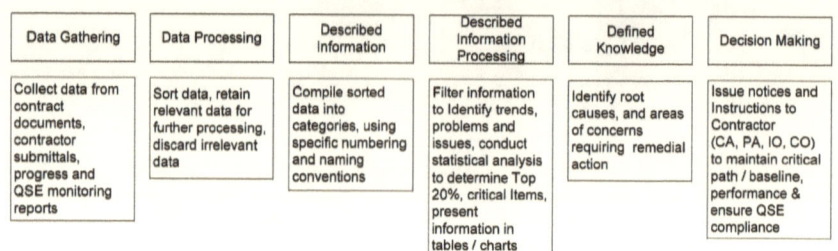

Figure 2.1 The six key steps of 6D.

There are key differences between data, information, and knowledge. Data is typically sourced from observations and measurements as undertaken by CS. Data is the raw material of information. Once data is compiled, collated, and interpreted, it is given meaning, which transforms it into information. The data-gathering process is as much about recording and reporting site activities as these occur as it is about measuring to ensure compliance. This is a process that requires observation and testing of the works to gain sufficient data and, thereby, information that confirms the right things are being done as specified.

The information generated may need to undergo many reviews for it to be correctly interpreted. As information is examined, interrogated, tested, analysed, applied, and confirmed, knowledge is created. Knowledge is the awareness or understanding of facts derived from the information processed and from the self through perception, experience, and learning. And with knowledge of what is occurring, project practitioners have the potential to make effective decisions that either maintain or improve construction performance.

*'Information is not knowledge. The world is drowning in information but is slow in the acquisition of knowledge. There is no substitute for knowledge' (W. Edwards Deming).*

It is knowledge that enables effective decision-making. Knowledge provides the necessary applied enlightenment that allows appropriate choices to be made, which must be made in a timely fashion. Knowledge, however, is time sensitive, wherein any delay in the use of acquired knowledge or information reduces both its relevance and value. Through time and inaction, the acquired knowledge can easily become irrelevant and revert to a fact, data.

Effective decision-making can be achieved through a six-step process known as 6D, or D cycle. This six-step process consists of data gathering (1D), data processing (2D), described information (3D), described-information processing (4D), defined knowledge (5D), and decision-making (6D). See diagram below.

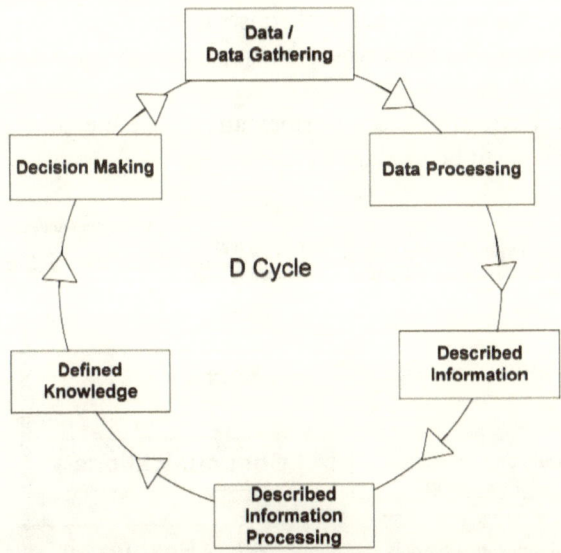

Figure 2.2 The 6D, or D-cycle process.

The decision-making process is, however, as important as how it is communicated. All decisions should therefore be communicated in a clear and understandable manner as instructions or notices issued by authorised persons. Equally, all instructions and notices should employ specific protocols, such as naming and numbering conventions so each element, activity, area, structure, zone, and project can be readily identified within the message and tracked.

## Performance Monitoring and Management

The project performance requirements in terms of QSE-STC (quality + HSE, scope, time, cost) for construction works are generally established by the client and articulated in the contract documents. To confirm, the CC should provide a work baseline schedule that details how the required performance for each item will be achieved. This baseline should include the following:

- quality, detailing how the quality (with acceptance and tolerance criteria) will be achieved
- HSE, detailing how health and safety and environmental protection compliance will be met
- scope, detailing the works to be undertaken with required processes and standards to be applied at each step
- schedule, proposed sequence of construction
- cost, outlining 4M resources to be employed, quantities, and measures

With these, both the CC's performance baseline and performance monitoring plan can be prepared.

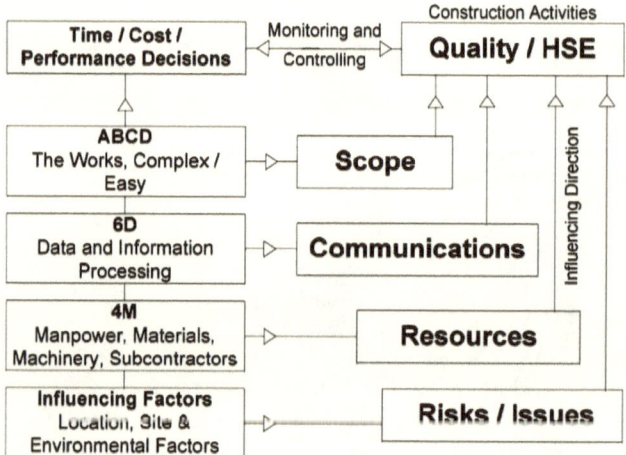

Figure 2.3 QSE and knowledge areas (STC-CHR), which can influence performance and the eventual outcome.

The performance monitoring plan must outline what data is to be captured, how it is to be processed, and what information is required in order to predict progress and make effective decisions that will improve performance (see chapter 5). Performance improvement decisions will in turn ensure QSE compliance as well as achieving the required outcomes in terms of time and cost.

# Chapter 3

# Phase 1: Pre-Construction—Initiating and Planning the Supervision Works

Prior to commencement of construction, it is important to determine and establish the required management processes. One key aspect to consider is the monitoring and verification of construction works—construction supervision. This is the only management process that can verify and, thereby, ensure the right things are done to the specified standard.

The choice to monitor and verify construction works will inevitably influence every other decision—from type of contract used, contractor selection, how site risks are to be managed and mitigated, how works are to be inspected and supervised, and how non-compliances are to be rectified. Equally, the type of contract adopted will influence the construction supervision process by the set contract conditions. It is therefore useful to review the contract, its requirements, and determine whether these are appropriate for the works prior to proceeding to construction.

Once the decision is made to supervise the works and appropriate terms are included in the construction contract, it is important to finalise the requirements. These requirements include:

- key appointment, roles, and responsibilities
- construction supervision process and requirements during various phase of construction

- pre-construction, pre-start requirements
- construction phase requirements
    o attendance at meetings
    o monitoring for QSE compliance
    o recording and reporting
    o issue of notices and instructions
    o reporting of incidents, accidents, and assisting in event management
    o performance monitoring.

- closing phase requirements, with respect to the following:
    o Cx (Commissioning) works
    o Cx checklists and notices
    o construction completion, handover, defect, and omission management.

All these aspects should be captured in a CS management plan. Once construction works starts, it is simply too hard to change the in-place processes. It is therefore vital to prepare and implement the CS's management plan prior to commencing construction work, preferably prior to executing the contract, so it can be progressively update with contract details and particulars prior to enacting. Once implemented, it then becomes a reference document for CS, who can use to the plan to ensure the works are performed in a safe manner, remain on time and on cost and are compliant without needing rework. Achieving this will also help ensure the desired planned project benefits are realised.

## The Construction Supervision Management Plan

*Plan to succeed; otherwise, you may be destined to fail.*

Project planning provides the means to ensure the required things are done at the required time so these achieve the required objectives and outcomes. The project management plan is the key document that should be prepared at the start of the project, produced to cover all knowledge areas processes and phases. As the project work proceeds, it is often essential to also prepare additional supplementary plans, such as the CS's management plan and Cx management plan, which are essential for construction works. Other essential plans that should already be implemented prior commencement of construction include the contract administration and project quality plans. Further information on these and other requirements can be found in the publication titled *The Project Manager's Checklist for Building Projects: Delivery Strategies & Processes* by the same author.

The management plan as a process tool that needs to be communicated to all project staff involved in the works for it to be understood and implemented. It is, therefore, essential that these plans sufficiently describe all required activities and, where appropriate, outline the actual requirements for each person and activity. And as required, updated regularly as work proceeds as this will create efficiencies and ensure resources are not unnecessarily wasted on activities that may not be relevant to the particular activity. Below is checklist for the CS management plan.

| Construction Supervision Management Plan Checklist | ✓/✗ |
|---|---|
| **Purpose** <br><br> Outline purpose of CS plan, include the following: <br><br> • key assumptions <br> • key inputs and essential references <br> • key appointments <br> • construction supervision phases and requirements. | |
| **Key Assumptions** <br><br> Confirm the CS process is based on the following key assumptions: <br><br> • The contract documents have captured the full scope of works that will satisfy client requirements and expectations. <br> • The design is fit for the purpose (unless the construction contract includes design activities), and no further design works are required. <br> • The CS is to ensure CC complies with QSE aspects only; other aspects, such as STC (scope, time, and cost) matters are to be monitored and managed by CAdmin and PM. | |
| **Key Inputs and Essential References** <br><br> List key inputs required to inform the construction supervision process, include the following: <br><br> • contract documents (ABCD) <br> • project management plan and supplementary plans, such as QSE management plan | |

| Construction Supervision Management Plan Checklist | ✓/✗ |
|---|---|
| • project governance structure and requirements<br>• certification requirements and statutory approval conditions<br>• company procedures and templates, quality management systems, supervision checklists, work instructions, HSE policy statements<br>• reference documents and relevant standards, such as OHS and environmental legislation, ISO 9000, ISO 14001, applicable laws, legislations, and statutory requirements<br>• CC pre-construction submittals, such as QSE management plans and related documentation, to ensure conformance with contract and anticipated site conditions. | |
| **Key Appointments**<br><br>List key appointments and management structure with respect to the supervision works, include the following:<br><br>• PM (project manager)<br>• CAdmin (contract administrator)<br>• CS (construction supervisor)<br>• CC (construction contractor).<br><br>All key appointments should be provided with an outline of their role and responsibility as specified in contract and as listed below. | |
| **PM (Project Manager)**<br><br>Outline PM's key roles and responsibilities with respect to construction supervision, include the following:<br><br>• Prepare and implement construction management plans for key knowledge areas—QSE, CS, and Cx management plans.<br>• Review CAdmin, CS, and CC performance.<br>• Oversee all aspects of the works, including the Q-STC-CHRIPS aspects, review compliance results, and manage performance to ensure works proceed as planned and in accordance with the set baseline, with required outcomes and outputs achieved.<br>• Brief client on progress, performance, and site concerns.<br>• Direct and oversee Q-STC change implementation.<br>• Ensure emergent or related issues are promptly resolved. | |

| Construction Supervision Management Plan Checklist | ✓/✗ |
|---|---|
| **CAdmin (Contract Administrator)**<br><br>Outline CAdmin's key roles and responsibilities with respect to construction supervision, include the following:<br><br>• Administer contractual aspects to ensure construction is in compliance with contract documents and in accordance with relevant regulations and applicable laws.<br>• Review CS reporting and performance to ensure site inspection are conducted as required.<br>• Review CC performance, monitor time (progress), cost (expenditure), quality (output), HSE performance, and issue notices (NC, CA, PA, IO).<br>• Communicate requirements to CC and results to PM.<br>• Implement authorised or required changes by issue of CO notice.<br>• Review CC's submittals to ensure compliance.<br>   - QSE management plan with all associated documentation—ITPs, shop drawings, RFIs, hazard assessment, SWMS, QSE reports<br>   - work (baseline) schedule with 4M information, invoicing forecast information, progress reports<br>   - construction work documentation submittals, as-built drawings, O&M manuals. | |
| **CS (Construction Supervisor)**<br><br>The CS's key role and responsibility is to ensure the QSE aspects are compliant with specified requirements during all phases of construction, which are as follows:<br><br>• phase 1: pre-construction phase<br>   - Review CC's pre-construction submittals (QSE plans and related documentation) for conformance with contract and anticipated site conditions in consultation with CAdmin and PM.<br><br>• phase 2: construction phase<br>   - Conduct inspections as required by contract and by nature of works, including the following: | |

| Construction Supervision Management Plan Checklist | ✓/✗ |
|---|---|
| <ul><li>o Undertake QSE monitoring of construction processes and outputs to ensure that CC fulfils all contractual requirements and statutory obligations.</li><li>o Conduct quality reviews of work undertaken, in progress, completed, and with materials and services provided.</li><li>o Undertake audits—process audit and QSE audits.</li></ul><ul><li>Review CC's construction submittals (RFIs, ITPs, shop drawings, commissioning plan) for conformance with established requirements.</li><li>Keep accurate records of work undertaken.</li><li>Report findings of conformances and non-conformances.</li><li>Monitor contract changes; ensure field- and engineering-directed changes are implemented as directed by CAdmin/PM (i.e. due to varying site conditions, such as underground obstructions or due to client change requests).</li><li>Conduct check survey of constructed structure to verify as-built information prior to acceptance of works.</li><li>Liaise with quality managers, building certifiers, and CxA (commissioning agents) to ensure all required works are inspected for compliance.</li></ul><ul><li>phase 3: closing phase<ul><li>Review CC final submittals—as-built drawings (including supplier and vendor drawings), operation and maintenance (O&M) manuals.</li><li>Facilitate the completion and handover process.</li><li>Ensure all defects are rectified and omissions completed prior to issue of final completion certificate.</li></ul></li></ul> | |
| **CC (Construction Contractor)**<br><br>CC's key roles and responsibilities include the following:<br><br><ul><li>Undertake construction works in compliance with contract documents.</li><li>Achieve QSE compliance by:</li></ul> | |

| Construction Supervision Management Plan Checklist | ✓/✗ |
|---|---|
| <ul><li>supervising work activity with contract requirements</li><li>producing works to the specified quality (and where not specified, in line with accepted industry standards)</li><li>maintaining a safe and healthy site environment by implementing safe work methods (controls) and by ensuring that all works are conducted in a manner that is safe and without risk to employees' health and safety or to the environment</li><li>implementing measures in accordance with QSE plan</li><li>managing construction risks identified from hazards assessment, applying controls and protection devices that eliminate, substitute, or administer to reduce hazards to acceptable levels</li><li>fulfilling all statutory requirements, ordinances, regulations, and by-laws of all authorities with jurisdiction over the works</li><li>taking appropriate and timely action to correct any deficiencies</li><li>investigating and reporting hazard and ensuring corrective actions are undertaken</li><li>preparing and participating in safety meetings and programmes, including conducting project inductions, toolbox talks, and daily team briefings</li><li>participating in accident or incident investigations</li><li>reporting all incidents to the proper authorities immediately and providing copy of the incident report to the CS, CAdmin, and PM in accordance with the related contract documents.</li></ul><br>• Document, record, and communicate work results.<ul><li>Maintain all construction work records—inspection reports, plant and equipment compliance, hazardous substance inspection, temporary structure compliance, changes to workforce and workplace.</li><li>Maintain induction and training records.</li><li>Report results and progress to CAdmin and PM.</li></ul> | |

| Construction Supervision Management Plan Checklist | ✓/✗ |
|---|---|
| • Effectively manage all 4M resources.<br>   - Provide training, information, and instructions to all employees (including for QSE aspects).<br>   - Ensure safe equipment and plant is provided and maintained. | |
| **CS's Requirements at Each Phase of Construction**<br><br>**Phase 1: Pre-Construction Phase**<br><br>The key items for CS to review and become aware of during the pre-construction phase include the following:<br><br>• project risk and hazard assessments and strategies to minimise risks and make works safe<br>• QSE monitoring plans (QC plan and HSE plan), with procedures and templates, covering the following:<br>   - supervision checklists and processes<br>   - recording and reporting processes—1D (data gathering), 2D (data processing), and 3D (described information)<br>   - NC (non-compliances), CA (corrective action), and PA (preventative action) processes<br>   - management of incidents, accidents, and events.<br>• CC pre-start submittals<br>   - organisational chart showing key appointments accompanied with position descriptions, authorities, summary of experience, and qualifications (include CC-PM, construction manager, QC and HSE managers)<br>   - contract submissions, such as certificates, bank guarantees (or security), construction management plan (incorporating site establishment plan, traffic management plans, QSE and industrial relations management plans), warranty deeds, and insurance coverage (to indemnify client, PM, CAdmin, CS against loss or expense incurred or claims made by third party arising from negligent act or omission by CC in undertaking the works) | |

| Construction Supervision Management Plan Checklist | ✓/✗ |
|---|---|
| <ul><li>procedures for managing identified issues and risks, detailing the following:<ul><li>management methodology to be applied</li><li>risk and issue register with action plans</li><li>schedule of reviews during works.</li></ul></li><li>QSE management plan (see CC-QSE Management Plan Submittal below)</li><li>list of subcontractor (ensure contract conditions are the same as main contract, and clauses include provisions for auditing, accident investigation, and removal of non-compliant workers)</li><li>site induction procedures</li><li>employee, subcontractor, and supplier training programme</li></ul><br>• schedule of required meetings<br>• others as required—completion requirements, Cx management plan, defect-and-omission management requirements, and final submittals. | |
| **Phase 2: The Construction Phase**<br><br>The key supervision requirements during the 'doing phase' include the following:<br><br>• Receive and review CC work submittals to ensure compliance.<ul><li>QC submittals<ul><li>ITPs, shop drawings, SWMS</li><li>certification for materials and products that require approval to be incorporated into the works</li><li>inspection requirements and frequency, RFIs</li><li>reporting of results and measures.</li></ul></li><li>HSE submittals<ul><li>construction works hazard assessment, hazard ID register</li><li>risk assessment with risk reduction, mitigation, and control strategies, applying principles of the hierarchy of control to minimise risk to personnel in the workplace, listing strategies from best to worst:</li></ul></li></ul> | |

| Construction Supervision Management Plan Checklist | ✓/✗ |
|---|---|
| <ul><li>eliminate, design out hazard</li><li>substitute, use less hazardous materials</li><li>isolate; use site rules, work permits, guards, barriers</li><li>engineer the solution, use materials not requiring site labour or maintenance</li><li>administratively establish systems of work to control risks and residual risks</li><li>apply training and use PPE and other safety measure.</li></ul><ul><li>work (baseline) schedule and invoicing forecast information</li><li>construction work documentation, including the following:</li><li>shop drawings, coordinated drawing detailing in large-scale layouts all disciplines (structural mechanical, electrical, plumbing, and fire protection) per location as required by contract</li><li>work-as-executed documents, as-built</li><li>product and material changes and proposed to be supplied and installed, with the following:<ul><li>manufacturer or vendor details</li><li>country of manufacture and the country of origin.</li></ul></li><li>Cx (commissioning) plan</li><li>4M information<ul><li>manpower (site staff CVs for QC, site foreman, contract administrator, and manager)</li><li>materials required for the woks, storage location, lifting, and movement within site</li><li>machinery to be utilised, maintenance details, traffic management plan</li><li>subcontractors to be employed.</li></ul></li><li>progress and QSE reports<ul><li>hazard incident reporting</li><li>safety record</li><li>QSE audits.</li></ul></li></ul> | |

| Construction Supervision Management Plan Checklist | ✓/✗ |
|---|---|

- Monitor site works and progress.
    - monitor site activity and works in accordance with ITPs, contract requirements, and as per schedule
    - monitor schedule and progress against baseline.

- Conduct regular site inspections to ensure QSE compliance.
    - Conduct QC to:
        o ensure quality procedures implemented
        o validate accuracy of test procedures
        o undertake audits as planned.
    - Monitor health and safety to:
        o oversee works to ensure OHS compliance
        o review site safety plan, evacuation and disaster plan
        o ensure CC conducts regular safety audits as work proceeds and promptly rectifies any identified unsafe situation
        o identify and report works where hazards are not managed, when unsafe work occurs, and to ensure contractor rectifies situation, reviews and investigation incident to prevent re-occurrence.
    - Monitor environmental aspects, ensuring agreed procedures are followed.
        o Ensure site management plan addresses all concerns associated with the site activities.
        o Waste, wastewater, solid waste, construction waste (for recycle, landfill), and hazardous waste management
        o water quality, water supply protection and storm water management
        o spill protection
        o fire protection
        o soil disturbance and erosion
        o stockpile management
        o air quality (odour, dust, and smoke)
        o noise control
        o vegetation
        o heritage
        o traffic management

| Construction Supervision Management Plan Checklist | ✓/✗ |
|---|---|
| <ul><li>o  emergency response management</li><li>o  site contamination, acid sulphate soils</li><li>o  food work hygiene (for on-site food services)</li><li>o  camp sanitation.</li></ul><ul><li>• Monitor implementation of agreed changes to the works.</li><li>• Record and report findings and results.</li></ul><ul><li>- Report progress against baseline.</li><li>- Keep daily site record at each site visit. The record should detail/provide the following:<ul><li>o  4M—manpower (number of persons on site), materials, machinery, and subcontractors</li><li>o  activities undertaken</li><li>o  weather conditions</li><li>o  traffic conditions and management</li><li>o  QSE assessment</li><li>o  photographs of works</li></ul></li><li>- Provide regular site activity reports, covering the following:<ul><li>o  progress summary, including assessment against baseline</li><li>o  summary of QSE aspects</li><li>o  quality aspects<ul><li>- corrective actions register, with documented evidence these were completed</li><li>- audits conducted</li><li>- lead indicators (e.g. how many inspections conducted).</li></ul></li><li>o  safety aspects, outlining the following:<ul><li>- incident reports, detailing notifiable incidents (e.g. also confirming relevant statutory authority have been contacted within the required period)</li><li>- lost time, medical treatment, and injury frequency rates</li><li>- number of visits from statutory authorities.</li></ul></li></ul></li></ul> | |

| Construction Supervision Management Plan Checklist | ✓/✗ |
|---|---|
| • Communicate as required with CC.<br>   - Respond to all requests.<br>   - Issue notices and instructions—NC, CA, PA and IOs.<br><br>• Arrange for confirmatory level survey of works to ensure compliance.<br>   - For multistorey structures, ensure benchmark plugs are installed to facilitate monitoring and to ensure correct levels, heights, and tolerances are achieved.<br>   - For civil works, locate benchmark plugs at prominent location—top culvert headwall, curb, abutment, at known locations, and accessible to level rod.<br>   - Provide geodetic elevations of established benchmark plug at completion of the works, and ensure these are noted on as-built drawings.<br><br>• Implement Cx management plan for plant and equipment.<br>   - Confirm operational requirements.<br>   - Appoint commissioning agent (CxA) to undertake pre-commissioning and commissioning works. | |
| **Phase 3: The Closing Phase**<br><br>As construction works nears completion, it is essential to have an established 'close out' procedure, which is to include the following:<br><br>• list of completion and handover activities<br>• monitoring regime to:<br>   - ensure control measures remain in place until risks and impacts are reduced to acceptable levels<br>   - ensure QSE compliance is maintained during the completion of the works until handed over to the client.<br><br>• defect-and-omission inspection to ensure rectification works continues until defects closed and works are confirmed as complete<br>• certification of works, occupancy certificate required prior to handover of works to the client for use<br>• performance evaluation of CC and contract works at completion. | |

The PM and CAdmin are responsible for preparing and implementing the CS management plan. Preferably, this plan should be issued to the CS for comment prior to commencing construction work. Once implemented, this plan should be strictly adhered to by the supervision team. And if it is found that the actual work processes differs substantially from what was planned, then only the PM and CAdmin should update the plan and issue new revision.

## Establish QSE Requirements

*Inspection is key to ensuring QSE compliance.*

The contract documents should require the person(s) undertaking the works, namely the CC, to be responsible for QSE compliance for all matters relating to the works and site. To achieve compliance, the CC may need to appoint suitably experienced personnel who are familiar with and understand the QSE requirements and are able to achieve the required performance.

*The aim is to have zero accidents or incidents during the works.*

A minimum performance requirement to comply should be established for all QSE aspects. The minimum requirements can be determined either by legislation or by contract (as specified within contract documents). Where possible, it is always preferable to refer to and apply best practice as this will ensure the desired QSE compliance is achieved safely without harm to the environment.

The following checklist can be utilised to establish QSE requirements.

| Establishing QSE Requirements Checklist | ✓/✗ |
|---|---|
| **Inputs and Requirements**<br><br>List relevant reference documents and requirements, include the following:<br><br>• contract and legislated requirements, which are to be captured in the following:<br>   - QSE compliance requirements<br>   - CC-QSE required submittal (to be submitted during pre-construction phase).<br><br>• specific industry requirements for particular sector (such as for oil and gas industry)<br>• standards<br>   - ISO 9001, quality standard, noting the following:<br>      o customer focus<br>      o employee involvement<br>      o process approach and system approach to management<br>      o continual improvement<br>      o fact-based decisions.<br>   - OHSAS 18001, occupation health and safety standard, noting requirement for organisations to:<br>      o minimise risk to employees<br>      o improve an existing health and safety management system<br>      o demonstrate diligence<br>      o ensure safe working conditions, product, and process.<br>   - ISO 14001, environmental management standard, noting the following:<br>      o compliance with applicable legal regulations<br>      o requirement to develop adequate policy and corporate procedures for environmental protection<br>      o control processes required to ensure environmental protection.<br><br>• company and personal obligations<br>   - contract obligations as defined within ABCD documents<br>   - statutory obligations as required by statute or legislation | |

| Establishing QSE Requirements Checklist | ✓/✗ |
|---|---|
| <ul><li>moral and ethical obligations as required to achieve a good practice standard.</li><li>risk management studies to assess identified risks for implementation of strategies, such as prevention, mitigation, and work around strategies required to avoid or contain occurrences</li></ul> | |
| **QSE Processes for Compliance**<br><br>Outline QSE compliance requirements, include the following:<br><ul><li>organisational<ul><li>roles and responsibilities for key appointments who has 'management and control' of the site and works.</li></ul></li><li>management<ul><li>management approach to undertaking oversight role</li><li>site meetings to review compliances at regular intervals and in accordance with ITPs and RFIs</li><li>recording and reporting of progress</li><li>rectifying non-compliances by issue of notices (NC, CA, PA, and IO)</li><li>process to confirm compliance prior to acceptance</li><li>regular reviews of assessed risks and mitigation measures implemented.</li></ul></li><li>monitoring and verification processes<ul><li>review against acceptance criteria and tolerance limits set for works</li><li>monitoring schedule, review of works for compliance at regular intervals and in accordance with ITPs and RFIs</li><li>process to resolve QSE issues that arise and non-compliance</li><li>processes adopted to enable performance improvement</li><li>audits scheduled at regular intervals.</li></ul></li><li>specific processes per phase<ul><li>phase 1: pre-construction phase<ul><li>Establish QSE obligations from legislation and contact requirements.</li></ul></li></ul></li></ul> | |

| Establishing QSE Requirements Checklist | ✓/✗ |
|---|---|
| <ul><li>o Prepare a register of QSE aspects requiring attention and impacts.<ul><li>Identify all activities requiring monitoring or which have, or may potentially have, a QSE impact.</li><li>Assess their impact by considering frequency or probability of occurrence and the severity of the impact.</li></ul></li><li>o Establish hazardous substances register and dangerous goods register.<ul><li>Obtain current MSDS (material safety data sheets) for all hazardous substances and dangerous goods.</li><li>Determine the risks of use, storage, handling, transport, disposal, and required procedures to control incident involving spillage or leakage.</li></ul></li><li>o Determine the environmental control measures to either prevent or minimise the identified environmental impacts.</li><li>o Establish monitoring schedule to achieve environmental compliance.</li><li>o Determine how these activities should be managed during the pre-construction, construction and closing phases.</li><li>o Agree with contractor the QSE requirements prior to granting possession of site.</li></ul> - phase 2: construction phase <ul><li>o Ensure compliance is achieved with all statutory and contract requirements.</li><li>o Oversee works in accordance with QC requirements, ITPs, RFIs, SWMS.</li><li>o Oversee 4M items, including the following:<ul><li>Manpower—PPE, training, safety programme, work permits, injury and damage reporting, work over or adjacent hazards, fire prevention, first-aid facilities, and adequate site facilities for the workforce</li></ul></li></ul> | |

| Establishing QSE Requirements Checklist | ✓/✗ |
|---|---|
| <ul><li>materials—haulage, delivery, storage, weight, size and type, disposal plan, MSDS (material safety data sheets) for hazardous substances</li><li>machinery—type, number, size, calibration, and maintenance requirements (with certification), tools (and portable tools), formwork, ladders, scaffolding, electrical and mechanical equipment, cranes and rigging equipment, plant operations, and on-site operations, transportation, excavations</li><li>subcontractors—work permits, contractor camps, and facilities (safe connection to services and utilities).</li></ul>- Cx process<br>   o Ensure compliance is achieved with Cx requirements of the built form.<ul><li>Include building envelope, lighting, air quality, ventilation, humidity.</li><li>Complete prior occupation as this can impact the health and well-being of those who are to occupy the spaces.</li></ul>- phase 3: closing phase<br>   o Hand over obligations and responsibilities to client or facility operators prior to occupancy. | |

Establishing the QSE requirements is an essential first step to determining the required monitoring regime and thereby enables effective oversight of construction works. At a minimum, the above lists the key items for CS to focus on, especially items considered high risk, where danger is imminent in regard to health and safety or where an environmental incident is likely. Depending on the works, these may highlight the necessity for the CS to be given the authority to issue directive orders as required in order to rectify urgent non-compliance QSE and possibly with authority to suspend parts of the works until the situation is remedied.

### Example of Environmental Waste Management

As construction activities generate significant quantities of waste, the CC should be required by contract, in addition to the legislated requirements, to implement strategies that reduce the environmental burden on society associated with waste disposal. The adopted strategies should aim to reduce disposal of waste by source reduction, reusing, and recycling of materials. To do so, the CC must produce a waste reduction management plan which should include the following:

- stated commitment to minimise waste disposal by company
- assigned responsibilities for implementation of waste reduction programme
- assessment of waste streams and identified opportunities
- list of actions to reduce waste such as:
  - reducing volume of waste going to landfills or incineration
  - reducing volume of new manufactured materials
  - diverting waste for reuse or recycling.
- list of monitoring, tracking, and verification measures to ensure implementation, compliance, and effectiveness
- performance improvement measures to reduce waste over time.

As per above example, once the specific QSE requirements are known and assessed, then it is essential for the CC to devise appropriate measures and incorporate these in the CC-QSE management plan. Then as works are undertaken, each step, each activity and task must be monitored and verified on a continual basis, comparing what is done with what is required to be done. Where differences occur, an assessment must be made to determine whether the works are to be accepted (is within criteria) or whether corrective action is required to achieve compliance.

### CC-QSE Management Plan Submittal

An essential pre-construction submittal required from the CC is the QSE management plan which is to outlines how the CC proposes to manage the QSE items of the works. Once the CC's submittal is received, then it must be reviewed to verify that the required processes are incorporates into the plan. Below is a checklist to review the CC-QSE management plan.

| CC-QSE Management Plan Submittal Checklist | ✓/✗ |
|---|---|
| **General Items**<br><br>The QSE management plan should include the following general information:<br><br>• name and title, with contractor and client details, contact title, and number<br>• version control information, date, revision no, issue number, amendment, approval<br>• file reference. | |
| **Content of QSE Management Plan**<br><br>The QSE management plan should include the following:<br><br>• organisational overview<br>    - organisational chart indicating roles and responsibilities and levels of authority (diagrammatic)<br>    - company QSE system, processes, and procedures to be applied.<br><br>• management, appointment of suitably experienced key personnel<br>    - responsible persons to manage and ensure implementation of QSE aspects and processes.<br><br>• QA/QC aspects<br>    - description of the works<br>    - methodology and processes to be adopted, including the following:<br>        o monitoring and control measures, noting testing, calibration of instrumentation, and equipment<br>        o requirements for 4M resources to be utilised<br>        o submittals to be issued, ITPs, RFIs, reports, and certification<br>        o non-compliance rectification processes. | |

| CC-QSE Management Plan Submittal Checklist | ✓/✗ |
|---|---|
| <ul><li>schedule and plan for supervision works (HWPS, ITPs)</li><li>auditing programme of works and systems in place (to evaluate effectiveness).</li></ul><ul><li>HSE aspects<ul><li>list of identified critical safety risks and hazards with mitigation strategies</li><li>safety and environmental performance monitoring proposal to ensure compliance, including meeting schedules, relevant checklists, inspection schedules, and audits to be conducted</li><li>procedure for investigating and reporting work incidents (including near misses)</li><li>non-compliance rectification processes</li><li>emergency procedure and responses</li><li>site-specific safety management plan detailing following:<ul><li>site safety plan, listing protocols to be adopted to ensure health and safety is preserved for all people on sites, whether they are directly or indirectly involved in the project</li><li>safe work systems and safety rules</li><li>permit to work (procedure)</li><li>equipment readiness review (procedure)</li><li>accident investigation and reporting (including any observed near misses)</li><li>risk assessment procedure, risk management, including measures to control risks</li><li>job safety analysis</li><li>HSE induction training and refresher safety training</li><li>safety orientation for all persons entering site to acquaint themselves on worksite conditions, actual and potential hazards, equipment, and practices to minimise accidents</li><li>staffing levels of qualified safety personnel assigned to the works to ensure full and consistent implementation of safety policy and rules</li><li>weekly safety meetings with required attendees</li></ul></li></ul></li></ul> | |

| CC-QSE Management Plan Submittal Checklist | ✓/✗ |
|---|---|
| <ul><li>o safety of workers and positive accident prevention measures to be implemented</li><li>o emergency plan and procedures</li><li>o schedule of proposed safety inspection/audits.</li></ul><br>- site-specific environmental management plan detailing following:<br><ul><li>o environmental management plan</li><li>o environmental control measures</li><li>o responsibilities and requirements</li><li>o processes such as implementation, non-conformance control, corrective and preventative action, issues resolution</li><li>o environmental management controls to be employed, such as emissions, water quality, noise level control measures.</li></ul> | |
| **Reference Documents and Processes**<br><br>The QSE management plan should make reference to documents and processes to be applied, as follows:<br><br>• induction and training processes<br><br>   - induction training, task training, and refresher safety training requirements, noting the following:<br>      o requirement for all personnel to attend general QSE induction training before entering site<br>      o requirement for all personnel to attend adequate site-specific induction and site-specific work activity<br>   - safety training before starting works<br>   - list of identified QSE training needs for management, supervisors, and personnel on-site<br>   - processes to ensure appropriate training is carried out<br>   - record-keeping of HSE training.<br><br>• site safety rules commonly in use, noting the following:<br><br>   - site-specific safety rules that are to apply to the particular site and procedures to be used whilst on site | |

| CC-QSE Management Plan Submittal Checklist | ✓/✗ |
|---|---|
| <ul><li>- accident and emergency procedures with location first-aid facilities</li><li>- PPE requirements when entering the site</li><li>- measures to be taken to protect the public</li><li>- security of access, ensuring only authorised entry, movement on exit of persons, vehicles, and equipment.</li></ul><br>• PPE requirements whilst undertaking works, such as safety helmets and safety footwear, which are worn by all persons on site<br>• traffic management plan as approved and required by relevant authority, detailing the following:<ul><li>- location, road names, road designation, speed limits</li><li>- description of activity</li><li>- work programme</li><li>- proposed restricted work hours</li><li>- traffic details, peak-hour flows</li><li>- proposed traffic management method (active, unattended, night) and measures, proposed speed restrictions</li><li>- contingency plans</li><li>- public notification</li><li>- personal safety</li><li>- on-site monitoring (attended, unattended, night, and other times)</li><li>- other information (delay calculations, issues, temporary speeds, layout diagrams showing traffic controllers, and signage).</li></ul><br>• safe work procedures, including the following:<ul><li>- protection of workers and public</li><li>- use of barricades, fencing and overhead protection</li><li>- work at heights (elevated work), to be undertaken in accordance with the relevant construction safety legislation, standards, and codes</li><li>- electrical work installations and equipment, ensuring this complies with construction and electrical safety legislation, regulations, standards, codes, and ensuring utility services are identified and are made safe before commencing work.</li></ul> | |

| CC-QSE Management Plan Submittal Checklist | ✓/✗ |
|---|---|
| <ul><li>incident management procedure and processes, including the following:<ul><li>preventative measures accompanied with process that prepare for, respond to, and recover from incidents (with response team to be on 24-hour call)</li><li>incident management communication procedures, ensuring the following:<ul><li>Procedure is posted and displayed throughout the site.</li><li>All persons are made aware of procedures and location of first aid facilities.</li></ul></li></ul></li><li>communication management plan, ensuring QSE requirements are communicated to all persons on site</li><li>contract documents and legislated requirements, noting acceptance criteria, inspection requirements, HWPS</li><li>company quality system, processes, and procedures (to be applied).</li></ul> | |

# Chapter 4

# Phase 2: The Construction Phase— Monitoring the Works

*Monitoring is an important task that can be used to gauge how the works are faring and progressing. When combined with data gathering and processing and by correctly interpreting the information gained, the opportunity exists to have a real understanding as to how work performance can be improved. After all, performance is that one important aspect that can determine not only the future viability of projects but also of organisations.*

This is the 'doing' phase of the works as described at the beginning of this book, the 'do' part of the 'start, plan, do, check, act, and end' cycle. This is also the phase where the management plans prepared in previous phase are implemented.

Monitoring is an essential part of this phase, the construction phase. Monitoring is required to gain sufficient information that can be used to assess site conditions, ensure compliance, and determine how the works are proceeding. The information gained can also be used to track risks, control the quality of activities, and gauge work performance. Construction monitoring and verifying activities are key functions performed by the CS.

## Monitoring for Compliance

*Check and ask 'How well are we doing?' Act on the information received.*

What is done on-site and how work is undertaken determine whether the results and outputs are compliant and in accordance with the contract requirements. The knowledge areas as described in previous chapters (the Q-STC, HR + HSE aspects) are all interlinked as shown in the figure below. Any issue affecting the quality or HSE aspects will greatly influence project time, cost, HR aspects, as well as the eventual outcome.

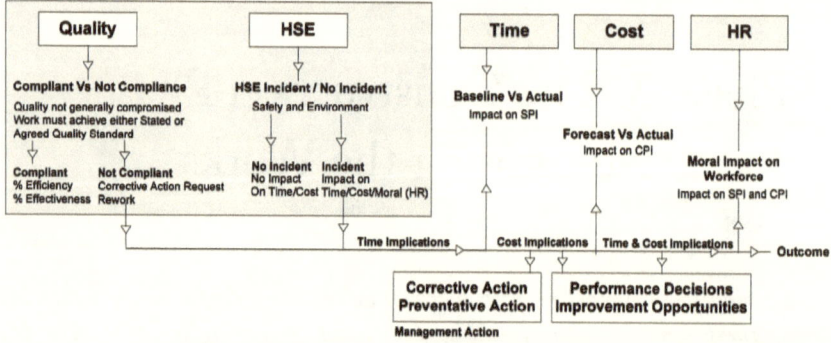

Figure 4.1 Links between Q-STC (quality, scope, time, cost), HR, and HSE.

How works are undertaken will determine whether it will be compliant or not-compliant and will determine whether further action will be required to achieve compliance and maintain planned schedule. The figure below outlines how QSE compliance is ensured. Monitoring and review of the works are essential to firstly identify non-compliances and, thereafter, ensure appropriate actions are applied to corrective any error, defect, or omission. This process will ensure the required performance maintained and possibly improved.

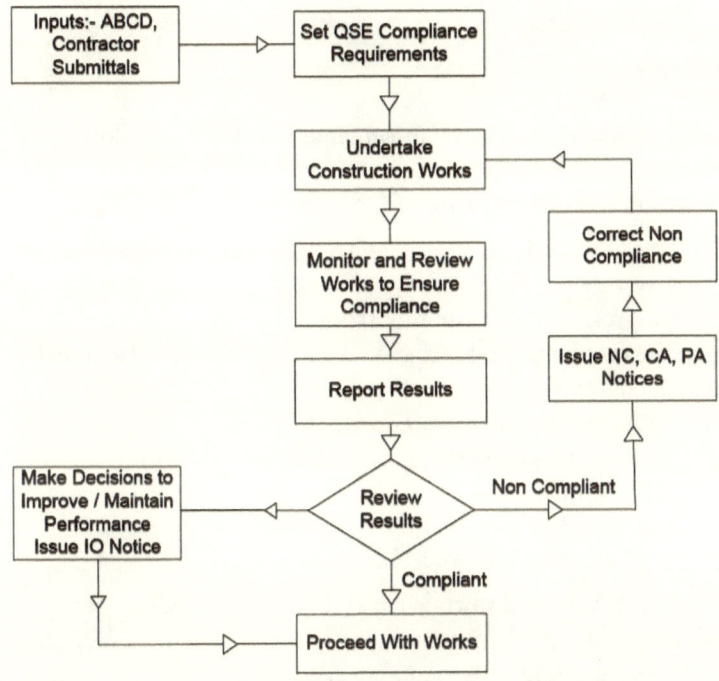

Figure 4.2 Monitoring, verification, and compliance process.

The construction supervision process provides the means to ensure the required things are done in the required way, in accordance with the contract, by identifying and correcting non-compliances as these occur. To achieve this, it is essential for the CS to have detailed understanding of the project requirements as described in the contract documents, together with a thorough understanding of QSE requirements to be applied.

Once the construction works commences, the CS must focus on monitoring the works to ensure these are compliant with the QSE requirements as stipulated within the contract documents. The inspection process can be aided with checklists of what should be inspected as per below.

| Inspection Checklist | ✓/✗ |
|---|---|
| **Item to Be Inspected** <br><br> For all types of construction works, both QSE and 4M activities and processes should be monitored to ensure compliance during construction phase. | |

| Inspection Checklist | ✓/✗ |
|---|---|

**QSE Items**

Ensure QSE requirements are implemented, include the following:

- Monitoring and reporting of QSE items.
    - Daily record of safety, with incident reporting and investigation, is current.
    - Environmental waste reporting is current.
    - Site meetings are conducted with QSE aspects discussed at each meeting.
    - Incident reporting procedure is in place.
    - Notices and instructions are issued to rectify non-compliance.

- Work procedures and processes.
    - CC SWMS submittals for all high-risk activities
    - ensuring QSE measures are incorporated into WMS (work method statements) or field procedures.

- CC's submittals, ITPs, RFIs, required inspections, and tests are scheduled and conducted as required.
- At the site.
    - Signs displaying the contractor's name and contact telephone numbers, including an afterhours emergency telephone number of the main contractor, are clearly visible from outside the site.
    - Where other contracts in place, the works are appropriately coordinated with other contractors.
    - Site inductions are conducted, and site registers, incident and accident registers are in place and current.
    - Site environmental management measures, as required, are in place and incorporate processes to:
        o create awareness of environmental aspects and impacts
        o minimise water and air pollution
        o minimise energy and water consumption

| **Inspection Checklist** | ✓/✗ |
|---|---|
| <ul><li>o apply effective and safe waste management, including hazardous wastes</li><li>o minimise potential impact on native flora and fauna.</li></ul><br>- Adequate facilities and equipment are available to contain, remove, and decontaminate affected areas in the event of an accidental spill or discharge of a pollutant.<br>- Emergency response procedure and evacuation plan is in place and rehearsed.<br>- First-aid station is provided and is staffed during hours of operation. | |
| **4M Items**<br><br>Verify 4M items are compliant with the contract QSE requirements, include the following:<br><br>• manpower<br>   - Verify manpower implemented as per contractor's schedule.<br>   - Ensure PPE, as per SWMS, are worn (safety helmets, boots, hearing protection, dust suppressed masks).<br>   - Ensure work is organised so it does not place workers in awkward or strained conditions.<br>   - Ensure dangerous works have SWMS, such as welding, requiring suitable screens are erected around area to prevent flash and sparks, and adequate ventilation is provided.<br>   - Ensure work is conducted safely.<br>      o For work at heights, monitor and ensure prevention of falls (guard rails, scaffolding, harness) in place, and all tools used are secured by lanyards.<br>      o For lasers, monitor works with laser beam, depending on the class of laser being used, ensure employees are trained and warning signs are displayed.<br>      o For confined spaces, monitor and ensure the use of portable ventilation equipment, safety harness, and appropriate barricades around manholes. At a minimum, two persons are present, and determine whether safety winch is required. | |

| Inspection Checklist | ✓/✗ |
|---|---|
| - For site amenities, ensure the following are provided:<br>   o First aid. Ensure first-aid box is on site, is fully stocked, and qualified first-aid person is on site.<br>   o Ablutions. Ensure appropriate amenities (1 toilet per 20 workers), with adequate fresh drinking water provided.<br>   o Cleanliness. Ensure site is regularly cleared of rubbish and scattered materials<br>   o Waste management. Ensure there is a designated rubbish disposal process area or waste container on site.<br>- For site safety, ensure that the following are provided:<br>   o Barriers. Ensure barricades and fences are erected along or around trenches and work areas.<br>   o Excavation. Ensure floor and roof penetrations are correctly managed and, if required, covered.<br>   o Services. Ensure all existing electrical, gas, and water service locations are verified.<br>   o Fire. Confirm site has adequate firefighting equipment available and equipment is serviceable.<br>   o Emergency plan. Ensure emergency and incident plans are rehearsed on a regular basis so all site personnel are conversant with procedure.<br>- Access, traffic, and safety.<br>   o Ensure warning signs are displayed.<br>   o Assess adequacy of public access diversion and site traffic.<br>• material<br>- Verify materials are delivered and installed as per schedule.<br>- Ensure all on-site hazardous substances (chemicals) have MSDA (Material safety Data Sheets) available on-site, are clearly labelled, and have instructions on label adhered to (stored in secure, ventilated spaces). | |

| Inspection Checklist | ✓/✗ |
|---|---|
| • machinery<br>    - Verify machinery implemented as per schedule.<br>    - Verify plant and machinery are in good order (no oil leaks, and hydraulic hoses are intact).<br>    - Ensure cranes have safe working load indicated on the boom.<br>    - Ensure vehicles have reverse alarm where necessary.<br>    - Ensure all operators have appropriate licences.<br>    - Ensure all machinery have valid registration or permit.<br>    - Verify safety check logbooks are in order for machinery.<br>    - Verify safety plan prior to allowing machinery work close to power lines.<br>    - Electrical safety.<br>        o For machines and tools, ensure these have guards fitted.<br>        o For leads and plugs, ensure condition is serviceable, extension leads are off the ground, not suspended above work areas, and not extended more than 5 metres without support.<br>        o Ensure plugs are not located in wet areas.<br>        o For equipment, verify these are regularly inspected and records kept of inspection.<br>    - Equipment safety.<br>        o Ensure all equipment is in good order (and regularly serviced).<br>        o Monitor ladder safety. Ensure these are fixed top and bottom and are no greater than 1 in 4 slope, are at least 1 metre above the top platform, and only timber or fibreglass ladders are used for work near electrical equipment.<br>        o Monitor scaffolding safety. Ensure these are secured against tipping (bracing in all directions, tied to building, base is stable), have secured deck, handrails, toe boards, and scaffolders are qualified. If the scaffold is under erection, ensure safety barrier is in place at base.<br>        o For explosive powered tools, ensure warning signs are in place, operator are licensed, PPE is worn (hearing and eye protection), toolbox is locked when not in use, and logbook is current. | |

The technical parts of the supervision process are discipline specific and, therefore, are not included in the above checklist. The value of this checklist lies in ensuring the QSE items are prioritised and given at least equal value to the discipline-specific parts of the works. Below is an example that integrates QSE (QC + HSE) with discipline-specific items for concrete works into one checklist.

**Example of Concrete Works Inspection Checklist**

Below is a list of items that required to be verified prior and monitored during concrete works.

- CC's submittals, SWMS, RFI for concrete works
- QC items
    - Check site works, confirm compaction of subgrade.
    - Check aggregates grade, trial mixes, test results.
    - Review shop drawings and engineering certification, check dimensions for compliance, measure quantities required before placement, and check rebar tensile strength, sizes, and placement for compliance.
    - Check preparations are complete, including coating applications, pre-wetting, sandblasting, water-jetting, or reinforcement.
    - Ensure all services, conducts, and voids are in place as required.
    - Ensure placement site is clean, ready for casting.
    - On delivery, ensure slump is checked by independent laboratory technician.
    - Monitor concrete pouring, ensure compaction of concrete by using vibrators, ensure spare vibrators remain in place during casting, ensure water–cement ratio is maintained, and confirm concrete is finished as required.
    - Monitor curing, ensure formwork is not removed during curing period, and ensure all specified treatment.
- HSE items
    - Ensure safety measures are in place and according to approved SWMS.
    - Ensure sufficient access is provided (remove obstructions) and in accordance with traffic management plan.
    - Review shop drawings and engineering certification.
    - Review engineering certification for formwork.

- Manpower
    - Ensure the number and experience of labourers proposed by CC are suitable for the works.

- Materials
    - Inspect formwork, and check materials for possible defects, mould releasing agent used (for ease in stripping), workmanship (for gaps which may lead to grout loss during casting), provision of chamfers, and water barriers where required.
    - Check batch plant proposed materials to be used for special concrete design mix requiring specific aggregate grade.
    - Confirm delivery, and note time, batch plant, concrete temperature. Visually check mix for its homogeneity and consistency.

- Machinery
    - Check equipment is suitable for works (wooden planks, vibrators, wheelbarrows).

- Subcontractor
    - Ensure subcontractors entering site are inducted and abide by all site rules.

## Communicating with CC

*Critical to the monitoring process is communications between parties.*

Communicating regularly with the CC is an essential part of the construction works. Communications should be both formal and informal; formal communications should include scheduled meetings, notices, and instructions, whilst informal should only be conducted to understand underlying facts and motives behind emerging issues. As issues arise, a regular and systematic approach that includes both formal and informal methods should be adopted to resolve significant issues, and once confirmed, decisions should be communicated in writing.

### Meetings

An essential part of the construction process is the communication of information. Essential to communicating relevant and timely information are face-to-face interactions as these allow the different parties to establish a relationship and discuss issues. To formalise the communication process, meetings should be held as these provide the means to disseminate key information quickly to key stakeholders, to discuss the work and the mechanics by which the contract is administered, to address specific issues as they arise, and to strategise how emergent issues are to be resolved. These should be held at intervals agreed between the parties, and are commonly either held weekly or monthly, depending on the work complexity.

It is normally the client's expectation that the CS/CAdmin will initiate and organise all necessary meetings with CC and with other stakeholders who have an interest in the project. One of the most essential meetings is the kick-off, pre-start meeting. This meeting should be held prior to the commencement of construction with all key stakeholders (PM, CAdmin, CS, CC, and others as required) to discuss and clarify contract requirements, expectations, levels of authority, and any specific issues or concerns.

The table below outlines the different types of meetings held during the construction phase with key requirements.

| Meeting Checklist | ✓/✗ |
|---|---|
| **Required Meetings and Attendees**<br><br>The meetings required to be held during the construction phase include the following:<br><br>• kick-off or pre-start meeting, which is to be attended by CC, CS, CAdmin, and PM<br>• weekly progress meetings, which are to be attended by CC, CS, and CAdmin<br>• coordination meetings, which are to be attended by CC, with subcontractors, CS, and CAdmin<br>• monthly progress meetings, which are to be attended by CC, CS, CAdmin, and PM<br>• quarterly progress meetings, which are to be attended by CC, CAdmin, PM, and client representative. | |

| Meeting Checklist | ✓/✗ |
|---|---|
| **Key Aim of Meetings** <br><br> Meetings are to be held to establish and maintain a close relationship between the key stakeholders to contract, with the aim to: <br><br> • facilitate the work progress and administration of the contract <br> • review works undertaken and identify areas requiring management action <br> • provide forum to raise concerns and issues for resolution <br> • provide historical record of discussions and resolutions by way of minutes of meeting. | |
| **Kick-off (Pre-Start) Meeting Agenda** <br><br> The first meeting with CC, which is to be chaired by PM, should be held prior to commencing the works and include the following: <br><br> • key contract requirements, outlining the following: <br>    - required performance, obligations, rights, and responsibilities of contracting parties <br>    - proposed schedule and plan for construction works <br>    - CC's submittal requirements, confirming that all pre-construction start requirements are met, including insurance for the works. <br><br> • administration procedures, such as: <br>    - progress payments <br>    - lines of communications <br>    - requirement for reporting, correspondence, and meetings <br>    - document control, coding structure for reporting, format, identifiers. <br><br> • the works-specific requirements, such as: <br>    - phases, sub-completion dates, completion date, and requirements for completion <br>    - site zones, building classifications <br>    - traffic management plan, access, site control, security, site safety <br>    - work schedule, work permits, and working hours for CC <br>    - required utility connection and services | |

| Meeting Checklist | ✓/✗ |
|---|---|

- property and right-of-way/access issues
- illumination for night-time activities, safety, and security
- survey requirements
- material sources, such as aggregate for roadworks
- specific-structure issues.

• work QSE requirements
  - appointment of QSE managers by CC and client
  - specific-compliance requirements with contract and applicable regulations or laws
  - QC requirements, confirm process for submitting and reviewing ITPs, SWMS
  - key safety items, such as:
    o safety requirements under contract
    o site safety rules
    o details of safety management plan
    o procedure for managing and reporting accidents
    o emergency procedures
  - environmental control requirements and issues
  - compliance checking
    o audits, schedule of QSE audits, and in-place system compliance audits (depends on contract, size, and risk profile of the work)
    o site inspection and monitoring (depends on contract, size, and risk of the work)
    o post-audit procedure to verify corrective action implemented.

• 4M aspects
  - manpower—names and contact details of key appointments and site responsibilities
  - materials—delivery and control
  - machinery—site access and traffic management.

• change management procedure for reviewing and allowing alterations or improvements to design or contract
• schedule frequency for future meetings and persons to attend.

| Meeting Checklist | ✓/✗ |
|---|---|
| - meetings with PM/CS/CC (site progress meetings)<br>- meetings with client<br>- CC site meetings, toolbox, and safety committee meetings. | |
| **Progress Meeting Agenda**<br><br>The following items should be discussed at each meeting held with CC:<br><br>• schedule, reviewing progress against baseline<br>• QC, reviewing the following:<br>  - results of site inspections<br>  - notices issued (NC, CA, PA, IO)<br>  - NC reoccurrences.<br><br>• HSE, reviewing the following:<br>  - safety issues<br>  - training conducted and required<br>  - serious safety breaches, accidents or dangerous occurrences, and rectification delays<br>  - repetition of previous breaches.<br><br>• CC's submittals, documentation, and updates, reviewing the following:<br>  - RFIs<br>  - SWMS for scheduled high-risk activities<br>  - shop drawings.<br><br>• 4M items—manpower, materials, machinery, subcontractors<br>• potential changes and change orders<br>• issues of concern relating to technical or QSE aspects. | |

All meetings should be recorded with action items noted and minutes distributed to all attendees, including the client, for reference.

**Recording and Reporting**

It is essential to keep complete and accurate records of all project information. As project information is recorded and communicated by way of reports, it is important to agree upfront on what is to be recorded, what data is to be captured, and the reporting requirements, the format, frequency, and distribution prior to commencing works. This will enable the required project information to be captured and reported in the similar fashion each time.

The table below outlines the general recording and reporting requirements.

| Reporting and Recording Checklist | ✓/✗ |
|---|---|
| **Recording Requirements**<br><br>Recording requirements should be specified to ensure consistency in reporting and in delivery of project information. The following should be adopted to record site activity:<br><br>• use of templates and checklists for:<br>    - prompting what should be captured<br>    - ensuring brevity, bullet-point information, to ease later transfer<br>    - issue of standard notices and instructions, such as NC, CA, PA, IO<br>    - recording test results, materials supplied, or work performed in the same format.<br><br>• use of photographs for:<br>    - highlighting issues of quality, improper methods used, safety or progress that may be in question<br>    - historical record, each photograph captioned with date, time, event, and details of occurrence (noting that these can become essential reference document that can be used to settle disputes of differences or claims)<br><br>• video recording for larger jobs for certain activities (provided that there are no associated industrial relations issues with recording workers without their prior approval). | |

| Reporting and Recording Checklist | ✓/✗ |
|---|---|
| **Reporting Requirements**<br><br>As works are undertaken, it is important to record and report what was monitored, what has occurred (accomplished tasks, milestones achieved), what is planned to occur (upcoming tasks), along with key issues and risks identified during the specific period. This information can be presented as follows:<br><br>• site activity reporting, which includes:<br>   - daily activity reports by CC<br>   - daily inspection reports by CS<br>   - inspection records with findings<br>   - diaries and daily records compiled into weekly reports<br>   - site log and record of site activity kept by both CS and CC's Foremen<br>   - photographs, recording site activities and progress, specifically where issues of quality, safety, or progress may be in question.<br><br>• QSE reporting, which includes:<br>   - inspections and audits undertaken, noting the following:<br>      o identified non-conformances<br>      o maintenance of records, review of records, reports, ensuring these are current.<br>   - hazard and risk assessments, register, and mitigation strategies<br>   - summary of QSE meetings held, such as toolbox meetings<br>   - compliance with environmental requirements, outlining the following:<br>      o approvals together with physical and procedural measures in place<br>      o confirming adopted measures are functioning as intended<br>   - incident and accident statistics with investigation findings and actions taken to correct and prevent reoccurrence. | |

| **Reporting and Recording Checklist** | ✓/✗ |
|---|---|
| <ul><li>4M reporting, which includes:<ul><li>manpower<ul><li>personnel—details of qualifications held by individuals</li><li>induction, training, and briefings given of site personnel—i.e. toolbox or equivalent meetings.</li></ul></li><li>materials<ul><li>list of client-supplied materials and CC acknowledgement of receipt when materials are delivered.</li></ul></li><li>machinery.<ul><li>safety equipment records.</li></ul></li></ul></li><li>site instructions, notices reporting, which includes:<ul><li>register of work instructions and notices issued (NC, CA, PA, IO)</li><li>summary of key issues</li><li>register of change notices issued</li><li>records of any complaints received and action taken.</li></ul></li><li>investigation reports, which includes:<ul><li>list of incidents, accidents, assessments, and CC claims</li><li>site record.</li></ul></li><li>work-as-executed reporting, including listing received as-built documentation</li><li>construction programme updates, which includes:<ul><li>comparisons of actual, with baseline, highlighting differences</li><li>schedule of key activities verified as complete.</li></ul></li><li>minutes of the meetings</li><li>project diary provided on a daily basis by CS with detailed description of work undertaken</li><li>weekly and monthly construction activity reporting outlining the following:</li></ul> | |

| Reporting and Recording Checklist | ✓/✗ |
|---|---|
| - what occurred, inspections undertaken, and what is scheduled to occur<br>- list of issues, notices, and correspondence.<br><br>• monthly performance assessment report (prepared by PM for issue to client). | |

The completed reports per period provide sufficient information that can be further assessed and processed to gauge overall project performance and, thereby, also enable effective decision-making. This is the essence of the 6D (D cycle), a process where data is transformed into useful information and effective decisions, as further explained in the next chapter.

**Example of Bridge Construction Reporting**

The required reports for the construction of a bridge should include the following:

- CC's submittals
    - shop and work-as-executed drawings
    - materials supplied
    - construction programme and schedule of activities
    - ITPs, SWMS, RFIs.

- CC's reports
    - minutes of the meetings
    - test results
        o pile driving report
        o concrete cylinder field report.
    - record of lateral stressing
    - record of concrete pour
    - biweekly timesheet, including extra work
    - daily gravel record, bulk material quantities, records
    - vehicle measurements
    - record of salvaged materials
    - scaffolding, propping, and formwork compliance reports
    - equipment rental agreement summary report.

- CS's reports
    - daily and weekly reports
    - site logbooks, work record forms
    - photographs
    - site memos.

- CAdmin reports
    - contract administration reports
    - investigation reports
    - site instructions and notices.

- PM's performance reports.

**Notices and Instructions**

Where it is not possible to rectify issues, bad workmanship, or concerns immediately on-site, a formal notice or instruction should be issued. For consistency, standard forms should be used for all written notices and instructions, including when communicating non-conformances and required corrective action.

Notices for issue should be reviewed against the listed information below.

| Notices Checklist | ✓/✗ |
|---|---|
| **General Requirements** <br><br> All notices and instructions as part of the contract should be issued formally, as follows: <br><br> • including contract details, brief description of work, contract number, and name of CC <br> • consecutively numbered, have an identifying number, numbered sequentially, registered, tracked with one copy filed, noting the following: <br>   - If an instruction is written but not issued, it should be cancelled and marked with the reason of non-issue. <br>   - If oral instruction is given, it should be confirmed in writing as soon as possible by the authorised person (CS/CAdmin). | |

| Notices Checklist | ✓/✗ |
|---|---|
| <ul><li>signed by authorised person and dated</li><li>categorised according to purpose, such as:<br>  - approval notice<br>  - permission-to-use notice<br>  - compliance certificate<br>  - NC (non-conformance) notice<br>  - CA (corrective action) notice<br>  - PA (preventative action) notice<br>  - notices issued by authorities.</li><li>including required completion date and space to note actual date of completion</li><li>including a signature space for CC to sign and put the date on the receipt.</li></ul> | |
| **Approval Notices**<br><br>Approval notices should only be issued by CAdmin or PM for contractual matters, responding to the following:<br><ul><li>change of project personnel</li><li>design change (after contract award)</li><li>change to scope of work</li><li>change to the construction schedule, including start and completion date</li><li>contract claim</li><li>extra work</li><li>contract over expenditure</li><li>acceptance of work for progress and final payment.</li></ul> | |
| **Acceptance Notice**<br><br>Acceptance notices should only be issued by CAdmin or PM to confirm compliance of following:<br><ul><li>items inspected or tested</li><li>work completed as part of contract.</li></ul> | |

| Notices Checklist | ✓/✗ |
|---|---|
| **Permission-to-Use Notice**<br><br>Permission-to-use notice is issued when it is essential to retain CC's liability for the works under review. This notice ensures there is no transfer of responsibility for items received, such as shop drawings, whilst allowing work to proceed. This notice is given for items reviewed and confirmed as acceptable for use, such as:<br><br>• QC and HSE plans<br>• SWMS<br>• ITP<br>• shop drawings. | |
| **Compliance Certificate**<br><br>A compliance certificate is issued as part of work assessed as complete for payment purposes. | |
| **NC (Non-Conformance) Notice**<br><br>A NC should not be issued for incomplete work or for problems relating to design. NC notice is issued when:<br><br>• a deficiency in characteristic, documentation, works, or procedure is identified from audits, surveillance, inspections, or other observations and assessed as unacceptable with respect to specified requirements<br>• an unsafe operating procedure is observed or has occurred<br>• a deficiency trend has occurred and similar work or an ongoing process, such as placement of concrete, is proceeding. | |
| **CA (Corrective Action) Notice**<br><br>Once NC is assessed and the likely cause of non-compliance is determined, a rectification or CA notice should be issued directing the following:<br><br>• specific party or person with responsibility to resolve the stated matter<br>• immediate remedial action required to resolve incident and mitigate damage | |

| **Notices Checklist** | ✓/✗ |
|---|---|
| • minimisation of possible adverse consequences by ensuring appropriate procedures and resources are applied until the matter is resolved. | |
| **PA (Preventative Action) Notice** <br><br> PA notices are issued where it is likely that an incident will occur or reoccur. <br><br> The required preventative action should be directed at rectifying root causes of the occurrences. | |
| **Notices Issued by Authorities** <br><br> Notices may be issued by authorities direct to CC for identified non-compliances after conducting site visits. <br><br> For instance, authorities responsible for HSE will conduct random site inspection and may issue notices direct to CC, who should be required (contractually) to immediately inform CAdmin and PM. | |

Once non-conformances are identified, these should be investigated and reported. In determining what action should to be taken, the following options should be considered:

- Apply corrective action to rectify non-conformance by issue of CA, PA, IO, or CO. This option should be implemented as soon as it is reasonably practicable after identification.
- Accept the non-conforming work or service with or without concessions (if minor) and, if required, amending ineffective procedures to facilitate future works.

### NC (Non-Conformance) Reporting

Once NC is identified, the NC should be assessed with findings recorded for reporting purposes and to support the issue of further notices (CA, PA). The NC report should be in the following format and provide the listed information below.

| NC Report (Example) | |
|---|---|
| NC No. | |
| Item Identifier | Inspector/Inspection Date |
| Inspection Criteria | Visual/Measurements/Laboratory test results or Quality plan/Specification and Standards |
| Case Description (when, where, who, what, etc.) | |
| Non-Conformance evaluation | Major/Minor |
| Root Cause | |
| Corrective/Preventative/Remedial Action CA/PA No. | Responsible signature |
| Corrective/Preventative/Remedial Action/Expected Finish Date | |
| Follow-Up, Verification Details/Date | |
| Close Out, CA/PA Completed | |
| Sign Off, Quality Manager/ Signature/Date | |

As NCs are documented, the focus should quickly shift to having these resolved. Immediate action is required to ensure the works are not needlessly delayed, and to achieve this, the following steps should be undertaken:

- Identify key problems and investigate root causes of these.
- Propose remedial actions, ensuring these are based on good information and analysis of causes, establishing solutions for each problem and root cause to avoid repeat.
- Document and communicate the solutions.
  - Assign responsibilities for action and setting the schedule to complete required work
  - Issue of CA or PA Notice to correct the issue.

- Review progress and follow up to ensure that actions and measures adopted are effective.

The number of issued NC is important as this will determine the extent rework required (CA) that may need to be undertaken (as this has associated time and costs) and whether preventive action (PA) will be required to avoid similar occurrences.

## CA (Corrective Action) Notice

*Immediate rectification notice should be issued for all safety matters.*

As key issues are identified as result of monitoring works, it is important to be clear what requires corrective action. Scope changes for instance should not be undertaken by issue of CA notice. CA should be issued to resolve immediate problems by:

- rectifying works by correcting identified non-conformances
- reworking, scrapping defective work or service, and redoing
- implementing compliance procedure where there has been a failure to implement, such as safety measures.

The CA notice should provide the following information.

| CA Notice (Example) | |
|---|---|
| CA Notice No. | |
| Request Source/NCR No. Item Identifier | |
| CA identified through (circle) | Audits/Non-Conform Product/ Client Compliant/Development Proposal/Injuries/Accidents |
| Case Description of Identified Hazard or situation requiring correction/rectification (when, where, who, what, etc.) | |
| Description of CA (consider hierarchy of controls, i.e. elimination, substitution, isolation, engineering controls, administrative controls, personal protective clothing, and equipment) | |
| Root Cause | |
| Risk Assessment, Possible Impact if not Actioned | |

| CA Notice (Example) | |
|---|---|
| CA Initiator/Name/Date | |
| Approval Name/Date | |
| Expected Date to Finish CA | |
| CA Follow-up Date | |
| Following up Date | CA Finish/CA Under Processing/ CA Not Yet Started |
| Close Out | CA is effective/CA is not Effective, Follow-up Action Required |
| Sign Off, Quality Manager/ Signature/Date | |

Immediate CA measures should be directed to rectify items identified that can potentially harm worker health and the environment. If the same or similar problems exist elsewhere, a PA notice may also be required to prevent the problem from recurring.

## PA (Preventive Action) Notice

*Problems are opportunities to improve. Preventing problems can be substantially cheaper than fixing problems after these occur.*

PA notices should be issued when the likelihood of NC is high and there is an opportunity to prevent problems occurring before undertaking works, especially when there are large potential consequences. Whilst CA is issued to rectify deviations from the set requirements, PA is issued to anticipate possible deviations, directed at rectifying identified root cause of potential non-conformances. For instance, if liquid spills have already occurred several times in a material transfer area, addressing the root cause will prevent future spill from occurring.

PA is generally issued for the following:

- to prevent a similar problem or event from reoccurring later in the project
- to change a resource, process, or procedure due to previous failure to meet acceptance criteria
- training to be undertaken for certain activities as necessary skills are lacking.

Preventive action should be specifically directed at controlling QSE (quality + HSE), scope, time, and costs aspects. However, any PA that leads to a change to the plan, baseline, or procedures should be considered as a change request, so it is reviewed prior and approved or rejected as part of the change control process.

The PA notice should provide the following information.

| PA Notice (Example) | |
| --- | --- |
| PA Notice No. | |
| Request Source/NC No. Item Identifier | |
| PA identified through (circle) | Audits/Non-Conform Product/ Client Compliant/Development Proposal/Injuries/Accidents |
| Case Description (when, where, who, what, etc.) | |
| Data or Information collection/ Client's Initiative/Development Proposal | |
| Request Date PA Initiator/Name | |
| Case Description Case Study Group Signatures | |
| Root Cause | |
| Description of Preventive Actions (action should prevent reoccurrence) | |
| Expected Date to Finish PA | |
| PA Follow-up Date | |
| Following up Date | PA Finish/PA Under Processing/PA Not Yet Started |
| Close Out | PA is effective/PA is not effective |
| Sign Off, Quality Manager/ Signature/Date | |

Knowing when to issue PA notice requires substantially more experience than calculation. The decision to issue must weigh the upfront cost of requesting the measure with both the possibility of not requiring the action and the possibility of later requiring work to be corrected through issue of CA notice.

Vital to implementing any action is communicating requirements with all affected parties, such as employees and contractors, and later ensuring all effected documents, procedures, and processes are updated accordingly. Training of key personnel may also be required to effect the change across the work area and ensure updated documents, procedures, and processes are used. Once implemented, the processes should then be monitored, tracked, and later assessed to determine the effectiveness of measures and whether any further action is required.

## IO (Improvement Opportunity) Notice

*IOs provide the opportunity to achieve continuous improvement.*

During construction works, it is important to keep a register of items that were done right, done wrong, and could of been done differently to achieve a better result for future reference. This register is the lessons-learnt register, which should also note improvement opportunities. Similarly, once CA or PA is applied, the outcome should then be further evaluated to determine whether there are any improvement opportunities. This is part of the QA, continuous improvement process, which requires performance to be continually evaluated for the purpose of identifying opportunities for improvement.

Once an IO is identified, it should be issued as a notice. The IO notice should provide the following information.

| IO Notice (Example) | |
|---|---|
| IO Registration No. Item Identifier | |
| Opportunity identified through (circle) | Audit/Non-conformance/ Suggestion/Complaint/Client request/Lesson Learnt |
| Case Description (when, where, who, what, etc.) | |
| Management Action/Possible Impact if not Actioned | |

| IO Notice (Example) | |
|---|---|
| Risk Assessment | |
| Improvement Details | Develop new procedure/Alter existing procedure/Amend Work Method Statement/Other |
| Implementation of IO by Action required by | |
| IO Initiator/Name Request Date | |
| IO Follow-up Date Description of Completed Actions | |
| Close Out By Date | |

The IO process is not meant be used for implementing additional requirements to the works or contract, such as best practice initiatives, as these may exceed the contract requirements. Additional requirements to the contract add cost, time, and scope to the works and, thereby, should be initiated through the CO (change order) process.

## CO (Change Order) Notice

Any action that has the potential to change the works, such as schedule, management plan, baseline, procedures, contract, or scope requires a formal change request to be issued. No changes to the contracted works should be allowed without an approved change request.

No matter how a project is planned, there will always be some type of change. However, prior to accepting the requirement for the change and issuing a CO, the following should be undertaken:

- evaluation of the possible impact of proposed changes on scope, time, cost, risk, quality, resources, and customer satisfaction
- re-evaluation of risks and reassessing HSE factors with proposed change
- determining additional monitoring requirement to verify implementation of proposed change.

Where necessary, root cause analysis should be undertaken to investigate the reason for the change and identify any associated opportunity for future improvement.

## RFI (Request for Inspection) Notice

The RFI notice is issued by the CC to the CAdmin or CS when specific works are ready to be inspected, usually in accordance with the contract-specified HWPS. The RFI should provide the following information.

| RFI Checklist | ✓/✗ |
|---|---|
| **General Items**<br><br>The RFI should include the following general information:<br><br>• RFI number<br>• reference to contract clause, specification section, ITP or SWMS or person requesting RFI<br>• request number<br>• name of CC and subcontractor (as applicable). | |
| **Details of Inspection**<br><br>Ensure the RFI has provided sufficient detail to undertake inspection, including:<br><br>• description of works to be inspected<br>• location item to be inspected<br>• date and time for requested inspection<br>• type of inspection requested<br>• declaration of compliance, stating all work requested for inspection has been reviewed for compliance with respect to the contract documents prior to the request being made, as confirmed with dated signature by CC authorised person. | |
| **Monitoring, Follow-Up, and Closing**<br><br>Follow up with inspection, recording the following:<br><br>• date and time of inspection conducted<br>• inspector name, with signature and date<br>• comments and instructions issued for non-compliance<br>• list of documents referred, with space to list attachments and inspection report. | |

## Notice and Instruction Register

A register for NC/CA/PA/RFI notices should be kept current during the construction phase. This register will enable the tracking of issued orders and ensure these are followed up and closed by required completion date. The register will also enable items to be categorised for quick identification and referencing in accordance with instruction type, causes, and impact. Below is a sample register for reference.

| Identifier/ No. | Notice/ Instruction | Description | (Cost and Time) Assessment/ impact | Follow-up Orders | Issue Date (YYMMDD) | Expected Finish Date | Actual Finish Date |
|---|---|---|---|---|---|---|---|
| RFI0025 | NCZ1B2C W0122 | Site Safety— work at heights | $13,400, 5 days lost | CAZ1B2 CW0122 Site Safety (Barriers) PPE | 140220 | 140225 | 140225 |
|  |  |  |  | PAZ1B2 CW0075 Training | 140220 | 140320 | 140405 |

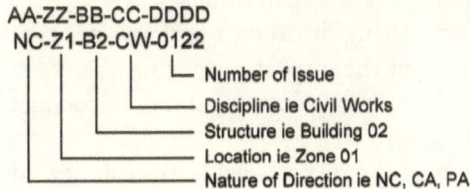

Figure 4.3 Notice Register with sample of numbering/naming convention.

In cases of recognised imminent danger or when the CC fails to comply with specific notices issued or fails to rectify previously identified worksite hazards that are regarded as extreme, the cessation of the work should be considered, and if ordered, work should not be allowed to resume until it is safe to do so.

## CC Construction Phase Work Submittals

An essential part of the contract is for the CC to submit certain documents to both inform what is intended and verify that the intended is in accordance with the contract. Once the CC's submittals are received, these must be reviewed to verify that the required processes, products, and quantities will be applied or installed.

Submittal documents during the construction phase can include ITPs (inspection and test plan), shop drawings, SWMSs (safe work method statements) and material data, samples, and product data. Submittal should be clear, readable, suitable for intended purpose, with separate issues for each items, referencing specification, location, and the works. Where material selections are proposed, submittal should include catalogues or samples from suitable manufacturers/suppliers as attachments.

## ITP (Inspection and Test Plan)

Generally, the 'what to test' and 'when to test' are determined prior to construction by considering the following:

- design requirements to verify critical items prior to cover up
- statutory requirements to ensure compliance with authority approval requirements
- project and client requirement to comply with QA, industrial relations, operations, and warranties conditions
- standard of contracting firms, past performance concerns
- cost of carrying out the inspection/testing
- type and complexity of work, accessibility to conduct inspections, and sampling
- management requirement to enable traceability, identification, and access to construction work information with aim to quickly determine problem areas and reduce non-compliant work
- risk and consequences of non-conforming works going undetected and failure occurring, considering the cost of remedial work, effect on construction programme, accessibility for rectification, disruption to use of building, consequential damage to other elements, threat to safety of workers and public.

Once determined, these can be specified in the contract documents with HWPs (hold and witness points) during the works, requiring the CC to provide ITP (inspection and test plan) for the stated items. HWPs are specific 'points' along the construction path requiring the CC to give notice to CAdmin/CS that the particular stage will be reached. A 'hold point' is where work must not proceed without an inspection or test verifying compliance, whereas a 'witness point' is optional, conducted where time permits, and does not require work to stop.

**Shop Drawings**

The CC should be required to issue shop drawings for items not fully detailed and requiring specialist input. Once received, these drawings should be reviewed for completeness and accuracy. The key review criteria are detailed in the checklist below.

| Shop Drawing Review Checklist | ✓/✗ |
|---|---|
| **General Items**<br><br>Shop drawings should be sufficiently referenced by include the following:<br><br>• site location of item or equipment within the project<br>• version control, with drawing and revision number and date<br>• item or equipment ID, detailing equipment, system summary and schematic, point-to-point wiring details, bill of materials, configuration details, construction details, as applicable<br>• certification certificate<br>• contractor's log, ID, with supplier and manufacturer details. | |
| **Compliance with Scope**<br><br>Verify compliance with scope by ensuring the following:<br><br>• All specified and scheduled items are included and exactly match requirements.<br>• Set-outs and dimensions are correct.<br>• Items are not substituted.<br>• Deviations are clearly identified and are assessed for compliance and acceptance. | |

| Shop Drawing Review Checklist | ✓/✗ |
|---|---|
| **Proposed Installation**<br><br>Review proposed installation, ensuring the following:<br><br>• Actual field conditions are represented.<br>• Proposed structure or equipment fits space as shown on construction and coordination drawings.<br>• Supports, erection, weights, and installation requirements are specified and does not void warranties or violate code requirements. | |
| **Equipment Items**<br><br>Review equipment proposed, ensuring the following:<br><br>• location of manufacture and origin noted<br>• control interface coordinated<br>• electrical characteristics listed (V/Ph/A). | |

## SWMS (Safe Work Method Statement)

The CC should be required to submit written SWMSs (safe work method statements) for all work activities assessed as having a high safety risk, such as work at height, with or near hazardous substances, in confined spaces, in deep excavations, and near electrical power transmissions. The SWMSs are an integral part of the risk and hazard assessment process as these are used to detail how the risks for specific works should be eliminated, minimised, or controlled.

The CC should be provided SWMS for each required activity for review prior to undertaking the works. It may also be necessary for the CC to submit progressive SWMSs, depending on the work activities and works programme. Once received, these must be reviewed by the CAdmin and CS for compliance and to ensure these are adequate and safe work can proceed. The key review criteria for SWMS are detailed in the checklist below.

| SWMS Review Checklist | ✓/✗ |
|---|---|
| **General Requirements**<br><br>The SWMS should be for a particular activity and in the following format:<br><br>• on CC's organisation letterhead, with name and registered office address of the organisation<br>• signed and dated by CC's senior management. | |
| **SWMS**<br><br>The SWMS should include:<br><br>• a description of the work to be undertaken<br>• the step-by-step sequence involved in doing the work (can be shown as flow chart).<br>• hazard assessment<br>    - potential hazards associated with the work and for each step of the work<br>    - safety control measures that will be in place to minimise these hazards.<br><br>• health and safety<br>    - details of all precautions to be taken to protect health and safety<br>    - details of all health and safety instructions to be given to persons involved with the work<br>    - identification and list of health and safety legislations, codes, standards applicable to the work, and where copies of these are kept.<br><br>• details of 4M items<br>    - manpower/subcontractors<br>        o Identity of all responsible persons, names and qualifications of those who will supervise the work, inspect, and approve work areas, work methods, protective measures, plant, equipment, and power tools. | |

| SWMS Review Checklist | ✓/✗ |
|---|---|
|     o  Details of training, description of what training is to be provided to people involved with the work, the names of those who will be or have been trained in the work activities as described in the SWMS, and the names and qualifications of those responsible for training them.<br>- materials, detailing storage and handling requirements<br>- machinery<br>    o  plant and equipment that will be used for the works, such as ladders, scaffolding, grinders, electrical leads, welding machines, fire extinguishers<br>    o  details of the inspection and maintenance checks that will be or have been carried out on the equipment listed. | |

From the issued SWMS, all key identified hazards and assigned responsibilities should be summarised in the table for ongoing tracking, management, and monitoring. Below is an example of hazard summary table for safety at heights work.

| Hazard Responsibility Table (Example) | | | |
|---|---|---|---|
| No. | Hazard ID | Strategy for Zero Incident/Accident | QSE Responsibilities |
| 29 | Safety at Heights Ref: SWMS 00343 | - Conduct reviews and assessment of site processes and compare with OHS legislative requirements<br>- Assess risk of injury<br>- Implement safety procedure, provide additional training<br>- Monitor for compliance. | - CC-HSE manager to submit SWMS<br>- CC to implementation and monitor compliance<br>- CS to oversee, confirm implementation. |

Figure 4.4 Example of hazard responsibilities table.

## Incidents, Accidents, and Events

*Prevention is better than cure.*

The role of the CS is to monitor and ensure HSE compliance; sometimes, however, even with effective monitoring and oversight, events do occur.

Events, in the most, are risks realised in the form of accidents or incidents. In these instances, the CS must ensure the CC takes immediate action to control the situation and manage the event as well as immediately informing the PM and CAdmin.

Realised risks are events. Event management is a reactive process requiring immediate action. If not controlled, these could easily escalate to disaster, where further and larger liabilities are incurred. Once the event is controlled or stabilised, only then the focus can be shifted to recovery. Unlike risk and event management, recovery requires preparedness to be effective. It, therefore, necessitates organisational commitment, where senior management must become actively involved as only senior management has the depth of experience, capability, authority, and ability to manage the many issues at play and ultimately determine the best course of action.

As realised events are brought under control, it is important that both the CS and CC record the details of event, with each conducting their own detailed investigation. It is also essential that all relevant facts (including HSE breaches) and occurrences are recorded as these can become crucial evidence when the matter is required to be recalled at a later date. The table below provides an example of an incident/accident report.

| Incident/Accident Report (Example) | | |
|---|---|---|
| **Item** | **Required Details** | **Details** |
| Incident/Accident | Date of Incident/Accident<br>Time of Incident/Accident<br>Investigation Date<br>Location/Workplace | |
| Persons Involved in the Investigation (Position/Name) | Name of person conducting investigation<br>Workplace Manager<br>Management HSE Nominee<br>Health and Safety Representative<br>Other: Name of person who was injured | |

| Incident/Accident Report (Example) | | |
|---|---|---|
| **Item** | **Required Details** | **Details** |
| Detail Injury Sustained/Damage Caused | | |
| Incident/Accident Location (State the exact location.) | Description of Incident (provide brief description) Has similar incident/near miss occurred previously? Were there procedures in place to minimise the risk? Has risk assessment for task been completed/reviewed (if applicable)? | |
| HSE (Medical or Environmental Treatment) Provide brief explanation of treatment measures and which authorities were contacted. | | |
| Environmental Key Contributing Factors (Provide a brief description of the circumstances that led to the incident/injury occurring and the immediate cause.) | Design of equipment/workplace (e.g. defective or unsuitable equipment, workplace layout) Environment (e.g. lighting, ventilation, noise, temperature) Human (e.g. fatigue, lack of understanding) Work methods and systems (e.g. training, unclear work procedures, flow of information) Other comments | |
| List of Documents Collected e.g. interviews, photos, safe work procedures and risk assessments | | |
| Corrective Actions (Provide detailed description of what actions to be taken to reduce the risk of the incident/injury from occurring again.) | Actions Completion Date Person Responsible Actions Completed | |
| Risk Management (Evaluate the likelihood, consequences and level of risk.) | Likelihood Consequence Risk Level | |

# Chapter 5

# Performance Management

*Performance is that one important aspect that can determine whether projects are on track, as well as the future viability of organisations.*

To understand how organisations are performing, their business, portfolios, programmes, projects, processes, people, and outcomes must all be monitored and assessed against pre-established targets and benchmarks, such as best practice, to determine whether these are in line with expectations and whether there are areas for improvement. With such information, organisations can determine what they are doing well and not so well, and implement strategies and tactics necessary that will enable them to remain viable for the longer term.

As projects are undertaken and progressed, it is essential to have an ongoing verification process of what was planned against what is being achieved. Simply having an 'unwavering' belief of success is insufficient as this can both be misguided and unfounded. Each step of the project together with each task and work must be continually monitored and verified with the set baseline, and where required, corrective action and improvement measures applied. Only then can it be ascertained that projects are proceeding as planned.

*You cannot manage what you cannot measure.*

It is essential to conduct periodic ongoing progress assessments as this provides a snapshot of how the project is faring and travelling. Such assessments enable the present position of project to be gauged and provide answers to the fundamental question of 'Are the required things being done?' The observed successes and failures can be used to 'feedforward' as to what immediate and future actions are also required. Similarly, comparing actual with best practice can provide insight as to the necessary improvements and possibly also help redirect the delivery strategy to achieve the required outcome more efficiently.

Simply observing how tasks and activities are undertaken and progress provides an insight as to the effectiveness of people, skills, and processes employed. As observations records are kept, each measure captures a snapshot of progress and achievement and, when assessed over a period of time, can provide the overall health of the work and, thereby, the project. As observational data is recorded, assessed, processed, and interpreted, it undergoes many phase changes, changing from data gathered to tabulated data/described information, to analysed information/defined knowledge, to decision-making and notices issued. This process describes how data flows to information and then decision-making, which is the basis for the 6D or the D Cycle as shown in figure below.

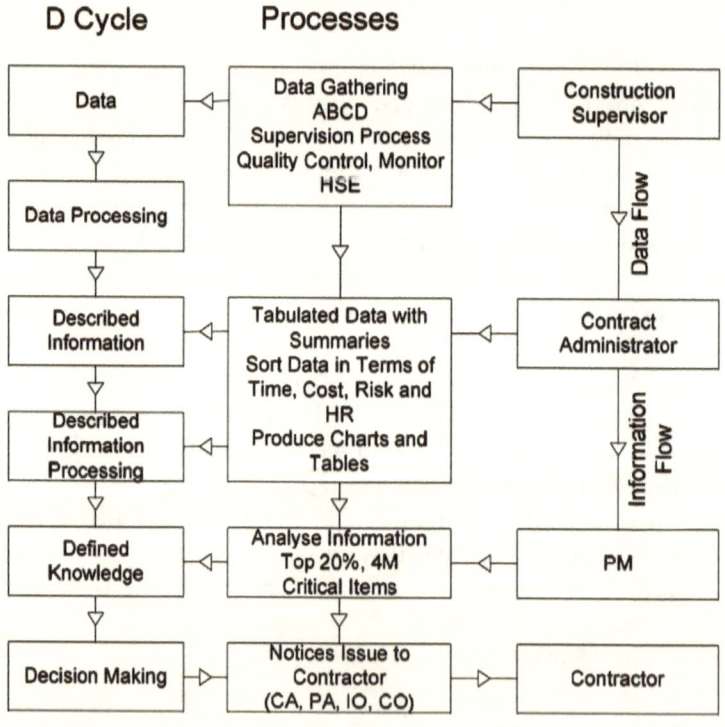

Figure 5.1 D cycle showing roles with data, information and decision flows.

Similarly, work and, thereby, project performance can be determined by monitoring and comparing actual results against the planned results, or the agreed baseline, in terms of cost, time, quality, and required deliverables and outputs. Assessing results provides an insight as to whether additional or remedial actions are required in order to fulfil the set contract requirements and thereby meet client expectations. With such insight, performance management is possible, where project decisions can be used to control what occurs. See figure below for how this relates to the 6D of the D cycle.

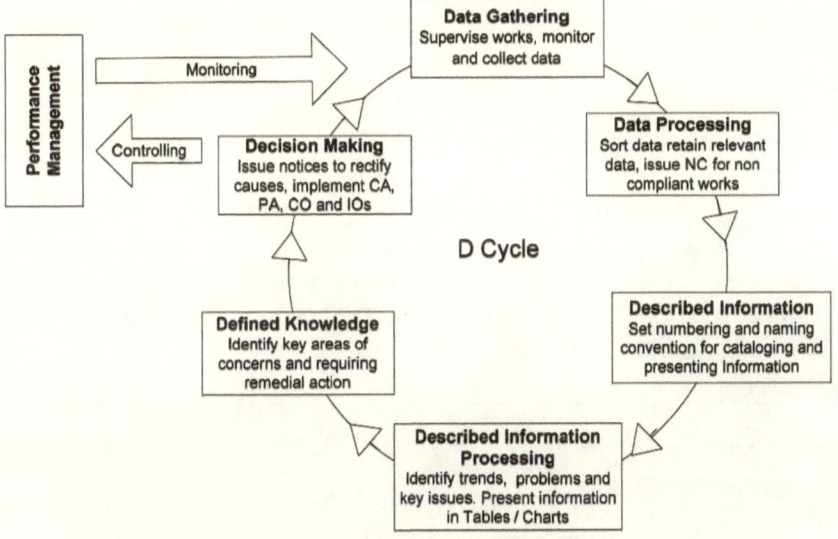

Figure 5.2 D cycle or 6D converting data from monitoring activities into performance decisions.

*Works can only be effectively managed (and controlled) when effective performance measures are employed.*

Performance should not be left to chance. To ensure the works produce to the required standard, the people must be given responsibilities, where individuals are made accountable for their actions and outputs. This is an essential part of managing the manpower aspect.

Responsibilities and accountabilities must be effectively communicated, articulated, and regularly monitored, with results given as frank feedback. Good performance can be rewarded, whilst shortfalls and areas requiring improvement followed up with corrective action rectification notices and with new expectations. Such feedback provides individuals with the means to focus, apply effort, and improve in line with work requirements. The table below outlines responsibilities and accountabilities with requirements for key project personnel.

| Construction Aspect | Responsibility to Monitor | Responsibility to Rectify | Performance Measures |
|---|---|---|---|
| Work Compliance/ Non-Compliance | Construction Supervisor | Contractor; reworks or additional works to correct omissions | SPI, CPI, % compliance, Rework, and Rectification Requirements |
| Project Progress | Construction Supervisor/ Contract Administrator | Contractor; reworks or additional works to correct omissions | SPI, CPI, Critical Path, and schedule variances from baseline |
| Project Time, Cost, QSE Targets, Critical Success Factors | Contract Administrator/ Project Manager | Contract Administrator/ Project Manager/ Contractor | % time, % cost, % QSE variances from baseline |

Figure 5.3 Performance measures with allocated responsibilities.

Legend: SPI denotes Schedule Performance Index (earned value / Present Value)
CPI denotes Cost Performance Index (earned value / Actual Cost)

As performance is measured and compared against the planned or desired performance, performance gaps can be rectified through effective decision-making and issue of notices and instructions. Effective decision-making requires appropriate knowledge, which must be gained through a process of gathering and retrieving relevant data, processing, and transforming it into useful and meaningful information that can be readily interpreted and applied. This process is outlined as the 6D or D Cycle, a six-step process that consists of data gathering (1D), data processing (2D), described information (3D), described-information processing (4D), defined knowledge (5D), and decision-making (6D), as further detailed below.

## 1D (Data Gathering)

The gathering of raw data is the first step of the D cycle. Data of what actually occurs must be captured form observation and monitoring activities. This raw data provides a historical record of what has occurred, and when processed, it can be used to assess performance and make decisions. It is of key importance to ensure that the data gathered is relevant, current, and sufficiently specific so it is of use.

Data gathering is a key activity for the CS, who must record progress, occurrences, results, and events during the works for all site activities. The key requirements and outputs of 1D (data gathering) are listed in the checklist below.

| 1D (Data Gathering) Checklist | ✓/✗ |
|---|---|
| **Key Inputs and Reference Documents**<br><br>Collect raw data from site activities with reference to the available documentation, include the following:<br><br>• supervision, monitoring, and QSE checklists, specifying what is to be captured<br>• contract documentation (ABCD) for specific monitoring requirements<br>• CC's submittals, RFIs, ITPs, SWMS, schedules<br>• monitoring reports and site activity logs. | |
| **Data Collection Requirements**<br><br>Outline what data is to be collected, include the following:<br><br>• works in progress or completed with respect to schedule<br>• QSE aspects, including the following:<br>   - compliance checking, certification, and commissioning works<br>   - quality audits<br>   - rectification works and reworks<br>   - incidents, accidents, events, and investigations<br>   - issue of NC, CA, PA, IO notices. | |

| 1D (Data Gathering) Checklist | ✓/✗ |
|---|---|
| **Outputs**<br><br>List the output data, include the following:<br><br>• CS reporting of QSE aspects, site inspections, work records, reports, notes, and diary entries<br>• CC weekly/monthly activity and summary reports of Q-STC and HSE aspects<br>• others, such as reports, summary of laboratory test results, CC's submittals. | |

Once raw data is collected from site activities, it can then be processed into a format that suits its intended audience and allows further processing for assessment. This is data processing or the second step of the D cycle.

## 2D (Data Processing)

2D is the second step in the D cycle. This step processes the gathered data in order to make it relevant and specific to certain work and areas of possible concern. The key requirements and outputs of 2D (data processing) are listed in the checklist below.

| 2D (Data Processing) Checklist | ✓/✗ |
|---|---|
| **Key Inputs**<br><br>Outline the required key information and reference documents from 1D (data gathering), include the following:<br><br>• contract documents, ABCD<br>• CC's submittals, including construction management plans, quality plan, QC procedures, HSE plan, ITPs, SWMS, schedules, WBS (work shown at activity level as work packages)<br>• QSE monitoring reports and results. | |
| **Processed Data Requirements**<br><br>Outline how data is to be processed, include the following:<br><br>• Sort to retain relevant data and discard irrelevant.<br>• Relevant data to be retained includes the following: | |

| 2D (Data Processing) Checklist | ✓/✗ |
|---|---|
|     - date of occurrence<br>    - ABCD recorded items, noting compliance<br>    - QC recorded items, noting compliance<br>    - HSE recorded items, noting compliance<br>    - 4M recorded items, summarising site activity.<br><br>• Data to be discarded as not relevant includes the following:<br>    - personal opinions that attack commercial reputation of CC<br>    - statements that may lead to legal action<br>    - statements that are understood to be opinions and not of fact<br>    - confidential information. | |
| **Outputs**<br><br>Outline how processed data is to be presented, include the following:<br><br>• QSE measure against baseline, noting the following:<br>    - variances<br>    - compliance and non-compliant works<br>    - corrective measures required.<br><br>• Issue NC for items identified as non-compliant. | |

Once data is processed, it is important to ensure that it is appropriately formatted so the information created is traceable and readily identified for further processing and evaluation. This leads to the third stage of the D cycle, where data is categorised and compiled into a useful format. This processing can ensure data inputted remains relevant, is traceable to point of origin, is easily identified, can be retrieved, and can be used to inform the performance assessment.

### 3D (Described Information)

Information is of value when it is appropriately referenced and compiled. Referencing enables the linking of what was planned (the design and contract documents) with the actual result (what is occurring) for each part of the works. Equally, referencing provides a readily adopted format that can be used to inform the decision-making process, making information accessible and easily retrievable.

The key requirements and outputs of 3D (described information) are listed in the checklist below.

| 3D (Described Information) Checklist | ✓/✗ |
|---|---|
| **Designator Requirements**<br><br>Outline numbering and naming convention to be used to catalogue and compile data into information, specifying the following:<br><br>• project or work designator<br>• location designator—location within site, zone, building, area, and within structure<br>• activity designators—linked to contract documents and WBS (work packages). | |
| **Outputs**<br><br>Provide designator format for the following:<br><br>• construction works at the activity level<br>• equipment to be supplied, installed, and commissioned. | |

To readily identify elements and, thereby, activities, most standard specifications adopt a specific numbering and naming convention that identifies specific elements of the works. MasterFormat[6] is one example of a commonly used standard specification that has a readily applicable and standardised way that organises information at the elemental level, which can also be applied to construction works. This specific numbering convention can be readily used to design, document requirements as it can be used to monitor and report on-site activity and construction works. When utilised, such a convention system has the potential to streamline both the communication and information management processes for the entire project. The figure below outlines how work site data can be processed so it becomes useful project information.

---

[6] The MasterFormat specification has a specific numbering and naming convention referred to as the Dewey Decimal System. This naming and numbering system has divisions and work sections which can be used to link design with the construction works. For example, designator 03 30 00 refers to division 03, concrete, section 30 00 titled 'cast in place concrete'. During planning, execution, and monitoring of construction works, such a numbering convention can be used to identify specific items of work, report these, and thereafter, undertake further analysis of the relevant work package.

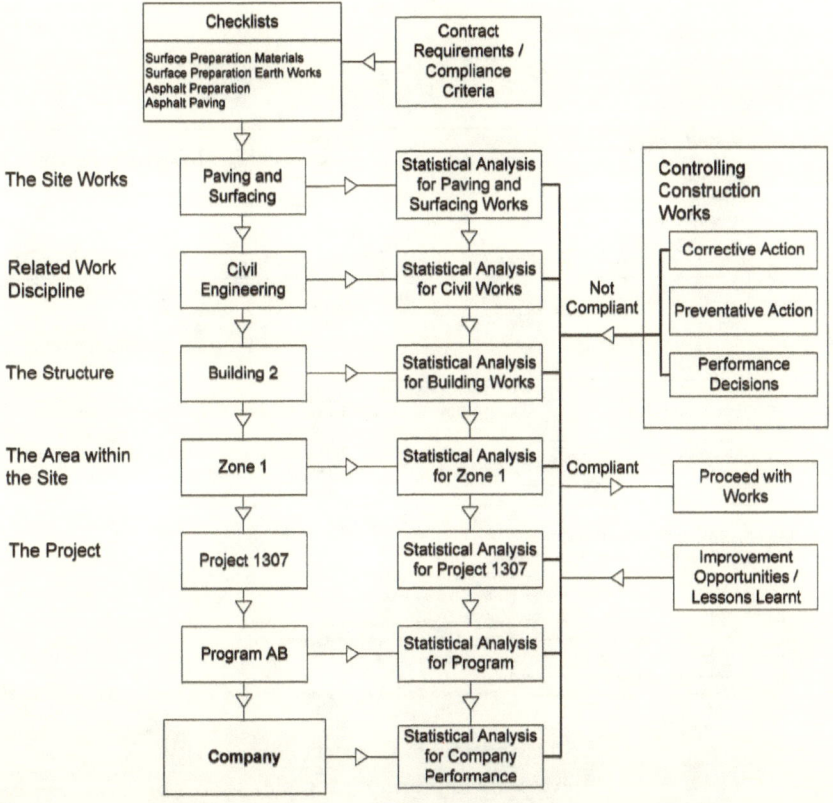

Figure 5.4 Outline of how data and information is described.

The use of element or activity numbering designators directly links contract documents with the site and HSE activities. The use of such designators also provides a base that can be expanded to include site and project information, which is essential to locate contractual item within the construction works. This also enables information to be systematically organised at the project level, allowing its categorisation in specific terms for ease of storage, retrieval, and late evaluation.

The tables below provide several examples of how the MasterFormat elemental numbering convention can be utilised as a base and expanded to include project activity or equipment information (as applicable) with relevant site location designators. Such conventions can be adopted project-wide and can be utilised by everyone when communicating, recording, and reporting.

| Designator | Designator Desciptor | Application | Reference / Note |
|---|---|---|---|
| AAAA or AAAAXX | A - Portfolio<br>A - Program<br>A - Project<br>A - Project Number<br>XX - Contractor Identifier | 1307 or 1307AB | Used to identify Project with / without Contractor |
| ZZ | Zone (for large Projects) | 1307-Z1 | Used when zones are designated |
| BB | Area or Structure | 1307-Z1-B2 | Used when there are 2 or more areas / structures<br>Area No eg A2 or Building No eg B2 |
| LLL | Level within Structure | 1307-Z1-B2-B04 | Building level number and or area according to stacking plan<br>eg B04 (Basement 4), L04 (Level 4), L00 (Ground Level) |
| WW | Discipline of Works | 1307-Z1-B2-B04 -CW | CW denotes Civil Works |
| CCCC-CCCC | Location within Structure | 1307-Z1-B2-B04 -CW -AA02BB02 | Building Grid Reference eg AA02-BB03, as provided on Construction Drawings |
| FFFFFF | Activity Level | 1307-Z1-B2-B04 -CW - AA02BB02-033000 | Full number refers to Specification Designator such as Master Format with first 2 letters denoting Division eg 03 for Concrete. |

```
AAAAXX-ZZ-BB-LLL-WW-CCCCCCCC-FFFFFF
1307AB -Z1 -B2 -B04 -CW- AA02BB02 - 033000
                          │    │      │    │         │      │         └─ Activity - Master Format Number - Cast in Place Concrete
                          │    │      │    │         │      └─ Building Grid Reference (from drawings)
                          │    │      │    │         └─ Discipline CW denotes Civil Works
                          │    │      │    └─ Building level number or area (ie basement 4)
                          │    │      └─ Area or Structure eg Building 2
                          │    └─ Zone
                          └─ Project No with Contractor / Supplier Identifier Number
```

Figure 5.5 Designators to Construction Activity Level for Monitoring (example).

| Designator | Designator Desciptor | Designator Example | Reportable Information | Responsibility |
|---|---|---|---|---|
| AAAAXX | Project Number with Contractor / Supplier designator | 1307-AB | Project Status / Contractor Performance | Program Manager |
| AAAA | Project Number (XX used to identify Contractor or Supplier) | 1307 | Project Status | |
| ZZ | Zone (for large Projects) | 1307-Z1 | Status per Zone | Construction Manager |
| BB | Area or Structure | 1307-Z1-B2 | Status for Area or Building | Quality Controller Manager |
| LLL | Level within Structure | 1307-Z1-B2-B04 | Status per Location within Structure | |
| WW | Discipline of Works | 1307-Z1-B2-B04-CW | Work Status for particulat Discipline | |
| | Discipline of Works (for general use without location specifier) | 1307-Z1-B2-CW | Discipline Status | |

Figure 5.6 Designators for Construction Reporting Purposes (example).

Construction Supervision: QC + HSE Management in Practice

| Designator | Designator Desciptor | Application | Reference / Note |
|---|---|---|---|
| EEE | Equipment Descriptor | AHU | Plant, System or Furniture Designator |
| ZZ | Zone (for large Projects) | AHU-Z1 | Used when project has zones designated |
| BB | Area or Structure | AHU-Z1-B2 | Used when there are 2 or more areas / structures<br>Area No eg A2 or Building No eg B2 |
| LLLLL | Location of Equipment within Structure | AHU-Z1-B2-01NW3 | Location of Equipment: Floor Level, Compass Location (NW, NE, SW, SE, N or S), Module Number |
| DDD | Equipment Description | AHU-Z1-B2-01NW3-SEN | Equipment Description, ie:<br>IMS – Information Management Server<br>SST – Sensor Suite<br>ADR – Air/Data Router<br>HFP – Vacuum Pump<br>OSC – MicroDuct Structured Cable<br>UPS – Uninterrupted Power Supply<br>XFM – Transformer<br>RS – Room Sensor<br>DPB – Duct Probe<br>SEN – Sensor |
| TT | Equipment Tag No. | AHU-Z1-B2-01NW3-SEN-68 | Tag No from 1 to 99 |
| MMM | Parameter Measured | AHU-Z1-B2-01NW3-SEN-69- CO, CO2 | List parameters measured, ie:<br>CO – Carbon Monoxide – ppm<br>CO2 – Carbon Dioxide – ppm<br>RH - Relative Humidity – percentage (%)<br>ENT – Enthalpy – BTU/lb<br>CRH – Calculated Humidity Ratio – grains/lb<br>DPT – Dewpoint – Degrees Celsius<br>TVC – Total Volatile Organic Compounds – ppm as isobutylene<br>PAR – Particle (PM 2.5) – particles/ ft3<br>T – Temperature - Degrees Fahrenheit<br>CON – Contamination signal - ppm as isobutylene<br>CVN – Contamination signal – Voltage |

EEE-ZZ-BB-LLLLL -DDD-TT-MMM
AHU-Z1-B2-01NW3-SEN-68-CO,CO2

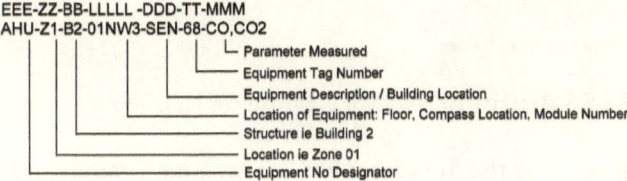

- Parameter Measured
- Equipment Tag Number
- Equipment Description / Building Location
- Location of Equipment: Floor, Compass Location, Module Number
- Structure ie Building 2
- Location ie Zone 01
- Equipment No Designator

Figure 5.7 Designators for Equipment to be Installed and Commissioned (example).

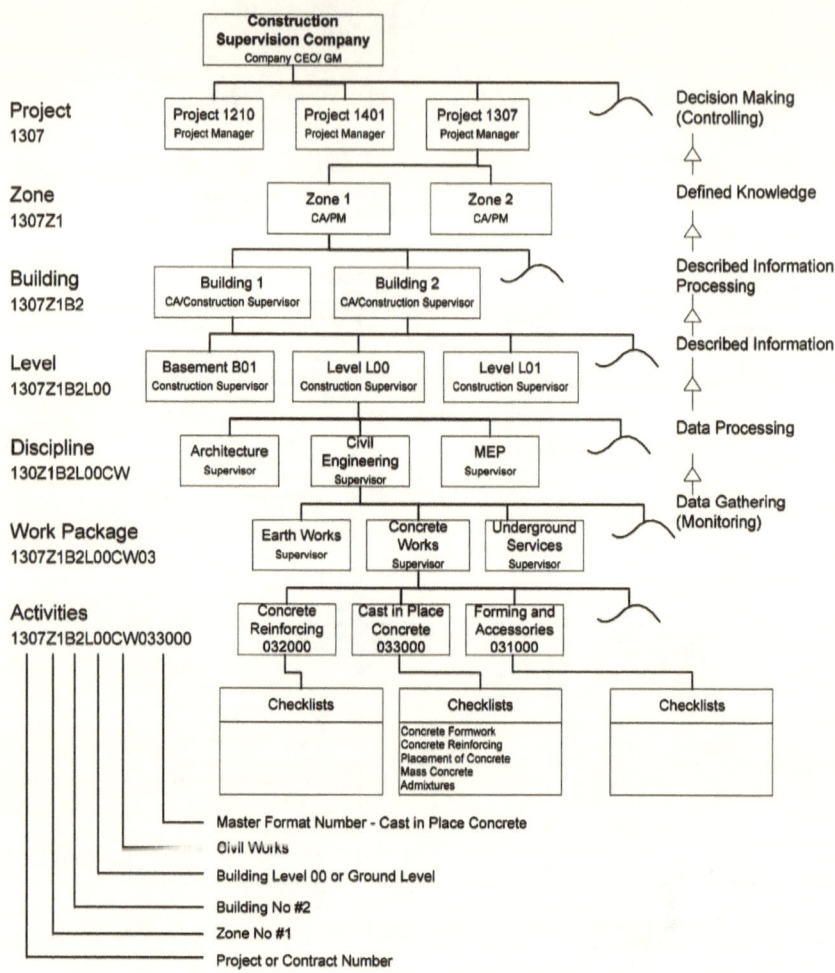

Figure 5.8 Activity Breakdown Chart (example).

Figure 5.8 summarises how the 3D (described information), numbering, and naming convention must categorise the work from the activity level to the structure/building, area/zone, and to the project level. It also outlines how decisions evolve from the data gathered at the activity level into project decisions.

The 3D (described information), in whichever format it is to be provided, should be structured so it can easily be interrogated and assessed and can allow the most critical items to be readily identified. This leads to the fourth stage of the D cycle, where information is further processed.

## 4D (Described-Information Processing)

Described-information processing is the process of further assessing available information. This is an essential step that must be undertaken in order to determine items of concern and critical items. The key inputs, requirements, and outputs of 4D (described-information processing) are listed in the checklist below.

| 4D (Described-Information Processing) Checklist | ✓/✗ |
|---|---|
| **Key Inputs**<br><br>Outline the required key inputs from 3D (described information), include the following:<br><br>• QSE information and measures against baseline<br>   - QC results, workmanship compliance information<br>   - HSE compliance information.<br><br>• 4M site information<br>   - manpower<br>   - materials<br>   - machinery<br>   - subcontractors. | |
| **Information Processing Requirements**<br><br>Outline the information processing requirements, include the following:<br><br>• List and sort issues according to type with previous notices issued (NC, CA, PA, IO, CO).<br>• Filter information to identify trends, problems, and issues.<br>• Conduct statistical analysis to determine critical/priority items and key issues and problems (Top 20 per cent). | |

| 4D (Described-Information Processing) Checklist | ✓/✗ |
|---|---|
| **Outputs** <br><br> Present processed information, include the following: <br><br> • QSE summary for each activity, structure, or building for specific area and zone and for the project <br>   - QC (number of non-conformances (NC) and reoccurrences) <br>   - HSE (number of HSE items, near misses, incidents, casualties, lost days) <br>   - list of critical items, top 20 per cent of issues and problems, root causes <br>   - cost of rework estimates <br>   - schedule delays, noting time to rectify non-compliance. <br><br> • diagrams, tables, and charts to highlight forecasts and trends. | |

The processed information should provide a QSE summary for each level of work, from the activity level to the structure/building, specific area and zone, and to how it relates to the project. To aid the assessment process, the information can be presented in diagrams, tables, and charts as these enable trends and areas of concern to be readily identified.

As the 3D (described information) is further processed, it is initially compiled into tables per discipline and for key elements and activities. From this list, it is possible to identify the 'stand out' or the top 20 per cent of the items—the key issues and problems. The top 20 per cent can be determined from the total issue of NCs, number of failures, incidents, time wasted, and in terms of additional cost.

According to the Pareto principle (or the 80/20 rule), 20 per cent of effort is responsible for 80 per cent of the result. With this rule, by identifying and rectifying the 'vital' few (20 per cent of defects), the 'many' (80 per cent of the problems) will be resolved. By focusing on the key problem areas, on what really matters, it is possible to determine what rectification works are required to achieve the greatest improvement. Ultimately, this has the potential to improve flows, remove root causes, provide uniform outputs, increase consistency and remove waste. Focusing time and energy on the 20 per cent enables the required things to be done at the required time.

Furthermore, each activity can also be classified in terms of their 'criticality'; in terms of their position within the critical path of the project schedule. From this assessment, it becomes evident where the key focus and effort should be placed; specifically on where the top 20 per cent identified key issues and problems coincide with items considered critical (for QSE example see figure 5.12 below). With further assessment, it is also possible to estimate the costs to recover[7] and maintain the initial baseline schedule (see figure 5.21).

Below are example tables showing 4D described information categorised and presented for the works, building, zone and project activities.

---

[7] Costs can include construction costs (rework) and supervision costs (as additional CS and CAdmin services may be required) to make the work compliant. It is important to note also that time wasted measures for each activity are only cumulative when these are on critical path and each impacts the project duration.

## Work Activity 4D Information Presentation Example

Project No.: 1307  Zone No.: Z1  Building No.: B2  Work Category: CW
Weekly / Monthly Report
From 22/02/2014 To 1/03/2014

| No. | Item Identifier | Master Format Reference | Item Description | No. of Inspections/Tests | No. of Confirmed/Accepted | No. of NCs/Failures | No. of HSE Incidents | Place on time Schedule | Estimated Waste in Terms of | | Corrective Action / Preventative Action / Incident Report |
|---|---|---|---|---|---|---|---|---|---|---|---|
| | | | | | | | | | Time(Hours) | Cost | |
| **Sub-Structure:** | | | | | | | | | | | |
| L00 | 1307Z1B2L00CW | 042200 | Concrete Unit Masonry | 5 | 4 | 1 | 2 | Critical | 2 | $10 | CA21B200033, HSEZ1B200012 |
| L00 | 1307Z1B2L00CW | 071100 | Dampproofing | 5 | 4 | 1 | | Critical | 2 | $20 | CA21B200034 |
| L00 | 1307Z1B2L00CW | 312300 | Excavation and Fill | 20 | 18 | 2 | 3 | Critical | 3 | $400 | CA21B200035, HSEZ1B200014, HSEZ1B200016 |
| L01 | 1307Z1B2L01CW | 042200 | Concrete Unit Masonry | 5 | 4 | 1 | | Critical | 2 | $10 | CA21B200033 |
| L01 | 1307Z1B2L01CW | 071100 | Dampproofing | 5 | 4 | 1 | | Critical | 2 | $20 | CA21B200034 |
| L01 | 1307Z1B2L01CW | 312300 | Excavation and Fill | 20 | 17 | 3 | | Critical | 3 | $100 | CA21B200036 |
| **Super-Structure:** | | | | | | | | | | | |
| L02 | 1307Z1B2L02CW | 042200 | Concrete Unit Masonry | 10 | 8 | 2 | | Non-Critical | 2 | $10 | CA21B200033 |
| L02 | 1307Z1B2L02CW | 071100 | Dampproofing | 5 | 4 | 1 | | Non-Critical | 2 | $20 | CA21B200034 |
| L03 | 1307Z1B2L03CW | 042200 | Concrete Unit Masonry | 10 | 8 | 2 | | Non-Critical | 2 | $10 | CA21B200033 |
| L03 | 1307Z1B2L03CW | 071100 | Dampproofing | 5 | 4 | 1 | | Non-Critical | 2 | $20 | CA21B200033 |
| L04 | 1307Z1B2L04CW | 033900 | Concrete Curing | 10 | 4 | 8 | | Non-Critical | 4 | $80 | CA21B200037, PAZ1B200002 |
| L05 | 1307Z1B2L05CW | 031100 | Concrete Forming | 4 | 3 | 1 | 7 | Critical | 8 | | CA21B200038, PAZ1B200002, HSEZ1B200013 |
| L05 | 1307Z1B2L05CW | 032100 | Reinforcing Steel | 2 | 1 | 1 | | Critical | 2 | $20 | CA21B200039 |
| | | | **Totals** | 106 | 83 | 25 | 12 | | 36 | $740 | |
| | | | **% Change** | | 78% | 24% | | | | | |
| | | | **Top 20% Margin** | | 5.0 | 2.4 | | | 7.2 | $148 | |

**SUMMARY**

| MS Reference | Civil Works | No. Inspections | No. Confirmed | No. NCs | No. HSE |
|---|---|---|---|---|---|
| DIVISION 02 | EXISTING CONDITIONS | 0 | 0 | 0 | 0 |
| DIVISION 03 | CONCRETE | 16 | 8 | 10 | 7 |
| DIVISION 04 | MASONRY | 30 | 24 | 6 | 2 |
| DIVISION 07 | THERMAL AND MOISTURE PROTECTION | 20 | 16 | 4 | 0 |
| DIVISION 31 | EARTHWORK | 40 | 35 | 5 | 3 |
| | Civil Works Totals | 106 | 83 | 25 | 12 |

Figure 5.9 Example of QSE summary table for civil work activities for Building B2.

| HSE Incident | | Corrective Action | | Preventative Action | |
|---|---|---|---|---|---|
| HSEZ1B200012 | Hard Hats not worn | CAZ1B200033 | Mortar Mix not compliant | PAZ1B200002 | Schedule and proceedure for concrete curing and form work |
| HSEZ1B200013 | Harnesses not worn for Work at Height | CAZ1B200034 | Sheet Waterproofing wrongly applied | PAZ1B200003 | Reinforcement Placement |
| HSEZ1B200014 | Water contamination with site waste | CAZ1B200035 | Backfill not failed compaction test | PAZ1B200004 | Admixtures to concrete |
| HSEZ1B200015 | Work commenced without Traffic Management Plan | CAZ1B200036 | Over-excavation | PAZ1B200005 | Traffic management approval process |
| HSEZ1B200016 | Stormwater Management | CAZ1B200037 | Formwork removed prior to required curing | PAZ1B200006 | Traffic management approval process |
| HSEZ1B200017 | Traffic Management | CAZ1B200038 | Certification of Formwork | PAZ1B200007 | Compaction testing proceedure |
| HSEZ1B200018 | Work permit for night time work | CAZ1B200039 | Incorrect Placement of Reinforcement | PAZ1B200008 | Stormwater management - prevention of flooding |

Figure 5.10 Example of HSE incident summary table for civil work activities for Building B2.

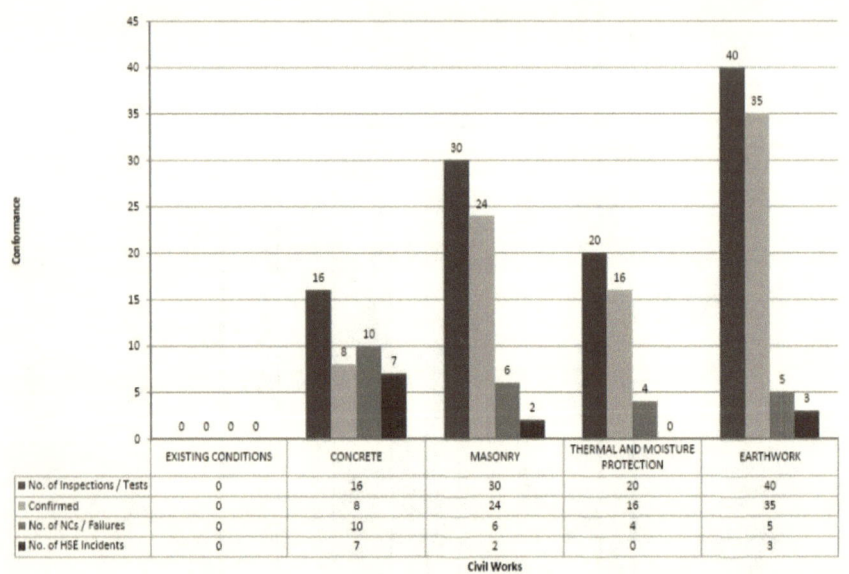

Figure 5.11 Example of civil works QSE summary graph for civil work activities for Building B2.

| Project No. : 1307 | | | | Zone No. : Z1 | | Building No. : | | B2 | |
|---|---|---|---|---|---|---|---|---|---|
| | | | | **Weekly / ~~Monthly~~ QSE Report** | | | | | |
| | | | | From 22/02/2014 To 1/03/2014 | | | | | |

| No. | Item Identifier | Item Description | No. of Inspections / Tests | No. of Confirmed / Accepted | No. of NCs / Failures | No. of HSE Incidents | Place on time Schedule | Estimated Waste in Terms of | |
|---|---|---|---|---|---|---|---|---|---|
| | | | | | | | | Time(Hours) | Cost |
| 1 | 1307Z1B2CW | Civil Works | 106 | 83 | 25 | 12 | Critical | 36 | $740 |
| 2 | 1307Z1B2AW | Arch. Works | 122 | 111 | 11 | 2 | Non-Critical | 48 | $500 |
| 3 | 1307Z1B2EW | Electrical Works | 75 | 70 | 5 | 1 | Non-Critical | 24 | $200 |
| 4 | 1307Z1B2MW | Mechanical Works | 20 | 18 | 2 | | Non-Critical | 20 | $100 |
| 5 | 1307Z1B2LW | Landscape Works | 10 | 8 | 2 | | Critical | 48 | $500 |
| 6 | 1307Z1B2SW | Special Works | 40 | 37 | 3 | | Critical | 24 | $250 |
| | | Totals | 373 | 327 | 48 | 15 | | 200 | $2,290 |
| | | % Difference | | 88% | 13% | | | | |
| | | | | Top 20% Margin | 10 | 3 | | 40 | $458 |

Figure 5.12 Example of QSE summary table for work activities for Building B2.

Construction Supervision: QC + HSE Management in Practice

**Weekly HSE Review Table**

Project No.: 1307  Zone No.: Z1  Building No.: B2

Weekly / ~~Monthly~~ Building HSE Report
From 22/02/2014 To 1/03/2014

| No. | HSE Item | Item Description | No of Incidents | Waste in Terms of | | HSE Rectification Instruction / PA |
|---|---|---|---|---|---|---|
| | | | | Time(Hours) | Cost | |
| 1 | PPE | Hard Hats not worn | 5 | 1 | $50 | HSEZ18200012, PAZ18200003 |
| 2 | PPE | Harnesses not worn for Work at Height | 3 | 28 | $300 | HSEZ18200013, PAZ18200003 |
| 3 | Waste | Water contamination with site waste | 2 | 24 | $200 | HSEZ18200014, PAZ18200005 |
| 4 | Excavation and Fill | Work commenced without Traffic Management Pl | 1 | | $50 | HSEZ18200015 |
| 5 | Stormwater | Stormwater Management | 3 | 20 | $200 | HSEZ18200016, PAZ18200004 |
| 6 | | | | | | |
| | | Totals | 14 | 73 | $800 | |
| | | Top 20% Margin | 2.8 | 14.6 | $160 | |

Figure 5.13 Example of HSE incident summary table for work activities for Building B2.

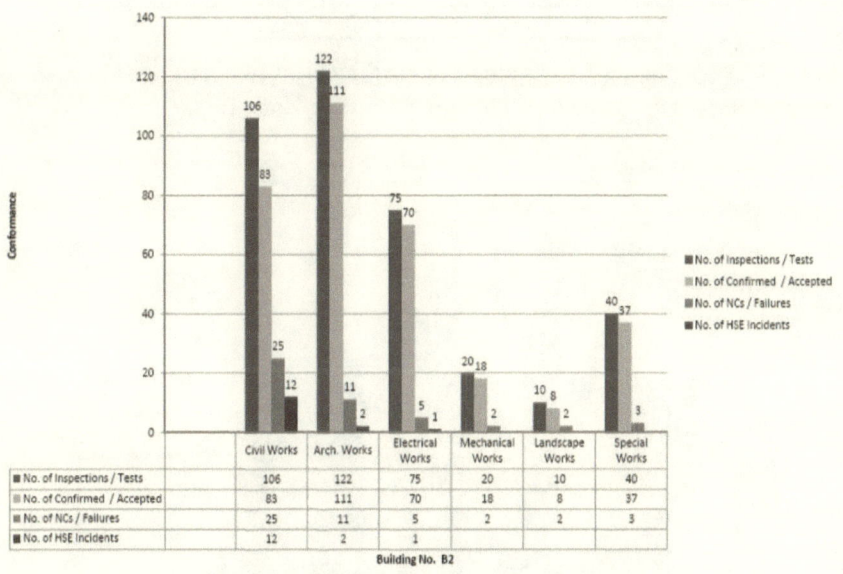

Figure 5.14 Example of QSE summary graph for work activities for Building B2.

# Building Activity 4D Information Presentation Example

Project No.: 1307  Zone No.: Z1
Weekly / ~~Monthly~~ Quality Conformance Status
From 22/02/2014 To 1/03/2014

| No. | Item Identifier | Item Description | No. of Inspections / Tests | No. of Confirmed / Accepted | No. of NCs / Failures | No. of HSE Incidents | Place on time Schedule Criticality | Estimated Waste in Terms of | |
|---|---|---|---|---|---|---|---|---|---|
| | | | | | | | | Time(Hours) | Cost |
| 1 | 1307Z1B1 | Main Building | 400 | 370 | 15 | 2 | Non-Critical | 12 | $1,600 |
| 2 | 1307Z1B2 | Adminstration Building | 373 | 327 | 48 | 15 | Critical | 40 | $2,290 |
| 3 | 1307Z1B3 | Utility Building | 220 | 200 | 20 | 3 | Critical | 13 | $1,500 |
| 4 | 1307Z1B4 | Resturant | 200 | 180 | 20 | 5 | Non-Critical | 15 | $1,600 |
| | | Totals | 1193 | 1077 | 103 | 25 | | 80 | $6,990 |
| | | % Difference | | 90.28% | 8.63% | | | | |
| | | Top 20% Margin | | | 20.6 | 5.0 | | 16.0 | $1,398 |

Figure 5.15 Example of QSE summary table for all buildings withing Zone 1.

Zone 01 Weekly HSE Review Table
Project No.: 1307  Zone No.: Z1
Weekly / ~~Monthly~~ HSE Report
From 22/02/2014 To 1/03/2014

| No. | HSE Item | Item Description | No of Incidents | Waste in Terms of | | HSE Rectification Instruction / PA |
|---|---|---|---|---|---|---|
| | | | | Time(Hours) | Cost | |
| 1 | PPE | Hard Hats not worn | 9 | 6 | $150 | HSEZ1B200012, PAZ1B200003, HSEZ1B300013, PAZ1B300003, HSEZ1B400007, PAZ1B400009 |
| 2 | PPE | Harnesses not worn for Work at Height | 5 | 28 | $300 | HSEZ1B200013, PAZ1B200003, HSEZ1B300014, PAZ1B300004, HSEZ1B400008, PAZ1B400010 |
| 3 | Waste | Water contamination with site waste | 2 | 24 | 200 | HSEZ1B200014, PAZ1B200005 |
| 4 | Excavation and Fill | Work commenced without Traffic Managem | 3 | 8 | $150 | HSEZ1B100015 |
| 5 | Stormwater | Stormwater Management | 4 | 20 | 200 | HSEZ1B200016, PAZ1B200004 |
| 6 | Electrical | Work commenced without work permit | 2 | 4 | $150 | HSEZ1B100012 |
| | | Totals | 25 | 90 | $1,150 | |
| | | Top 20% Margin | 5.0 | 18.0 | $230 | |

Figure 5.16 Example of HSE incident summary table for all buildings withing Zone 1.

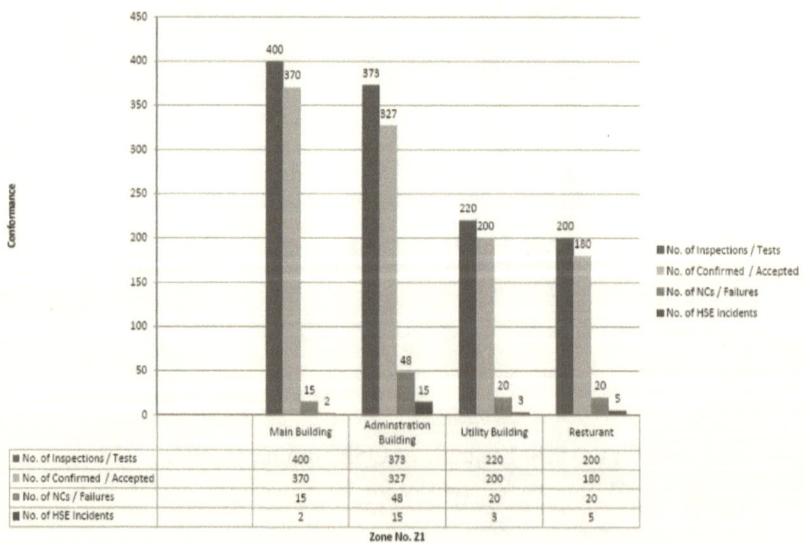

Figure 5.17 Example of QSE summary graph for all buildings withing Zone 1.

## Location Activity 4D Information Presentation Example

Project No. : 1307

From 22/02/2014 To 1/03/2014

| No. | Item Identifier | Item Description | No. of Inspections / Tests | No. of Confirmed / Accepted | No. of NCs / Failures | No. of HSE Incidents | Place on time Schedule Criticality | Estimated Waste in Terms of | |
|---|---|---|---|---|---|---|---|---|---|
| | | | | | | | | Time(Hours) | Cost |
| 1 | 1307Z1 | Zone No. ( 1 ) | 1193 | 1077 | 103 | 25 | Critical | 80 | $6,990 |
| 2 | 130722 | Zone No. ( 2 ) | 1850 | 1800 | 50 | 6 | Non-Critical | 10 | $2,000 |
| 3 | 130723 | Zone No. ( 3 ) | 1725 | 1700 | 25 | 11 | Critical | 13 | $1,500 |
| 4 | 130724 | Zone No. ( 4 ) | 1360 | 1350 | 10 | 10 | Non-Critical | 12 | $1,800 |
| | | Totals | 6128 | 5927 | 188 | 52 | | 115 | $12,290 |
| | | % Difference | | 96.7% | 3.1% | | | | |
| | | | | | Top 20% Margin | 37.6 | 10.4 | 23 | $2,458 |

Figure 5.18 Example of QSE summary table for all zones for Project 1307.

Project 1307 Weekly HSE Review Table

Weekly / Monthly HSE Report
From 22/02/2014 To 1/03/2014

| No. | HSE Item | Item Description | No of Incidents | Waste in Terms of | | HSE Rectification Instruction |
|---|---|---|---|---|---|---|
| | | | | Time(Hours) | Cost | |
| 1 | PPE | Hard Hats not worn | 30 | 7 | $300 | HSE000012 |
| 2 | PPE | Harnesses not worn for Work at Height | 6 | 60 | $600 | HSE000013, PAHSE000013 |
| 3 | Waste | Water contamination with site waste | 10 | 33 | $400 | HSE000014, PAHSE000014 |
| 4 | Excavation and Fill | Work commenced without Traffic Management | 2 | 8 | $300 | HSE000015 |
| 5 | Excavation | Traffic Management | 4 | 24 | $1,300 | HSE000033, PAHSE000033 |
| 6 | Site Induction | Failure to induct workers | 6 | 5 | $400 | HSE000018 |
| | | Totals | 58 | 137 | $3,300 | |
| | | Top 20% Margin | 11.6 | 27.4 | $660 | |

Figure 5.19 Example of HSE weekly summary table for all zones for Project 1307.

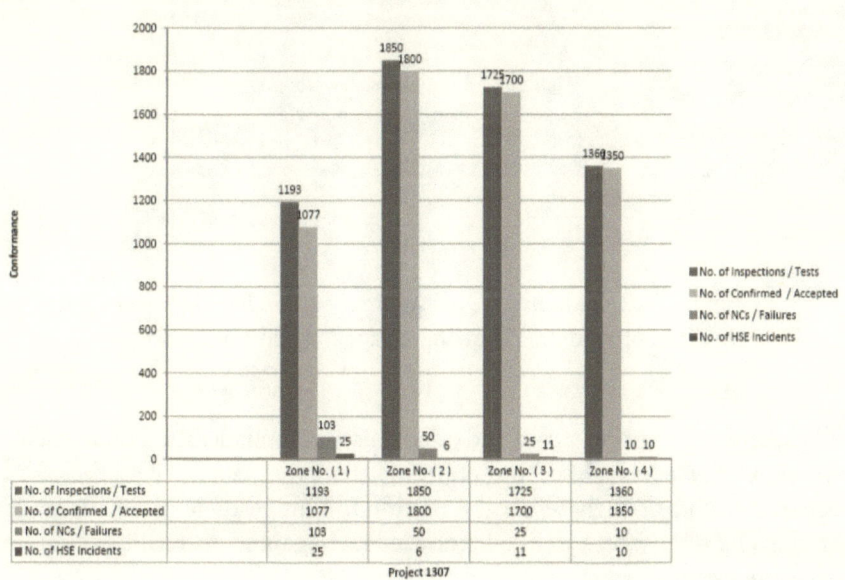

Figure 5.20 Example of QSE summary graph for all zones for Project 1307.

## Project Activity 4D Information Presentation Example

*Project* Activity 4D Information Presentation Example

| No. | Project No. | Project Name | Project Duration (Years) | Project Value $M | Place on time Schedule Criticality | Delay (days) | Added Cost $/month | Penalty Per Day | Penalties to Date | Estimated Cost to maintain Baseline | Action Summary (4M) |
|---|---|---|---|---|---|---|---|---|---|---|---|
| 1 | 1107 | XYZ Project | 2 | $35 | Critical | 12 | $25,000 | $1,000 | $11,500 | $20,000 | **Manpower** - Mobilise Additional Teams for 2 weeks<br>**Machine** - Mobilise additional Excavators for 2 weeks<br>**Subcontractors** - Mobilse Road Contractor within 1 week |
| 2 | 1401 | MNO Project | 2.5 | $20 | Non-Critical | 5 | $11,000 | $750 | $3,750 | $12,000 | Contractor accepted Delay Penalties |
| 3 | 1210 | AAA Project | 3 | $40 | Critical | 5 | $11,000 | $1,000 | $5,000 | $5,000 | **Manpower** - Mobilise Additional Teams and work double shifts for 2 weeks |
| 4 | 1309 | BBB Project | 2 | $15 | Non-Critical | 4.3 | $9,000 | $750 | $3,225 | $4,000 | Contractor accepted Delay Penalties |

Figure 5.21 Example of QSE summary table for all projects within programme of works.

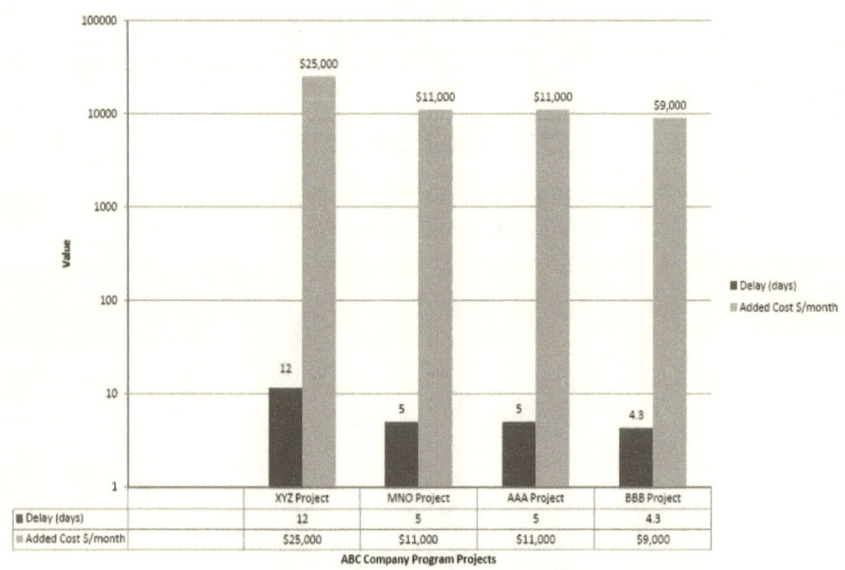

Figure 5.22 Example of QSE summary graph for all projects within programme, showing delays and additional costs.

The identified top 20 per cent and critical non-compliances and deficiencies should be further analysed to determine their root causes, why these are occurring, and how these can be rectified. This leads to the fifth stage of the D cycle, where defined knowledge is gained to enable effective decision-making.

## 5D (Defined Knowledge)

*How should results to be assessed? What is to be regarded as acceptable and as not acceptable? When should preventative action be applied?*

Defined knowledge is gained by correctly interpreting the available information, determining why certain events occurred, namely the top 20 per cent of non-compliances on critical path. This requires root cause analysis, a process which uses certain tools and techniques to identify and assess contributing and influencing factors that impacted the works and performance. Once assessed and determined, both corrective and preventive action can be applied to remedy the situation and prevent similar occurrences. The key inputs, processes, and outputs of 5D (defined knowledge) are listed in the checklist below.

| 5D (Defined Knowledge) Checklist | ✓/✗ |
|---|---|
| **Key Inputs**<br><br>Outline the required key inputs from 4D (described information), include the following:<br><br>• trend information<br>• critical and priority items<br>• key issues and problems (Top 20 per cent)<br>• percentage of change from baseline (critical items identified). | |
| **Define How Information Is to Be Processed**<br><br>Outline how information is to be processed, include the following:<br><br>• Identify and analyse trends, using the following:<br>  - check sheets, tally sheets used to gather data in a particular organised way for ease of further analysis<br>  - Pareto diagrams—vertical bar charts of frequencies or consequences used to identify the vital few sources (top 20 per cent) responsible for causing majority of problem (80 per cent)<br>  - histograms—bar charts used to describe the central tendency, dispersion, and shape of a statistical distribution | |

| 5D (Defined Knowledge) Checklist | ✓/✗ |
|---|---|
| <ul><li>control charts—used to determine the stability of process over time and whether it has predictable performance</li><li>scatter diagrams—correlation charts used to plot output variances in order to establish correlation between dependent variables.</li></ul><br>• note scope, schedule, and cost variances with respect to baseline<br>• determine root cause for areas of concerns and requiring remedial action | |
| **Root Cause Analysis**<br><br>List process tools and techniques to be used to assess critical items to determine root causes, why certain events occurred, include the following:<br><br>• cause and effect—process used to trace problem to its source and to identify root cause; fishbone diagrams are commonly used to link undesirable effects to the assignable cause upon and thereby also determine require action to eliminate cause<br>• tree diagram—systematic diagram used to represent decomposition hierarchies, such as the WBS, RBS (risk breakdown structure), and OBS (organisational breakdown structure), showing relationships links. | |
| **Outputs**<br><br>List identified root causes, include the following:<br><br>• management or procedure factors<br>    - faulty or missing procedures<br>    - poor communication<br>    - lack of understanding of requirements<br>    - failure to enforce rules.<br><br>• 4M factions<br>    - manpower (lack of training)<br>    - materials (defective products) | |

Construction Supervision: QC + HSE Management in Practice

| 5D (Defined Knowledge) Checklist | ✓/✗ |
|---|---|
| - machinery (malfunction or lack of maintenance)<br>- subcontractors (efficiency and effective measures of contractor, and workers on-site).<br><br>• instructions and notices (where corrective actions failed to address root cause)<br>• other factors: weather (where bad weather delays works). | |

Trend information can be obtained by simply determining the pass rate for inspections conducted. For deficiencies and non-compliances, the criticality of the item can be determined by assessing the issue of NCs with rework required that are on critical path. The additional 'work' may impact the works by increasing the overall schedule time and project costs. For the CC, this can translate to loss of profits, and for the CS, this can translate to additional work hours required to complete task and works.

Prior to completing the 5D (defined knowledge), it is important to ensure the data and information is accurate and remains relevant (is current) as this will affect the next stage, 6D (decision-making), influencing the type of decisions made.

## 6D (Decision-Making)

*Decisions are largely influenced by available information and past assessments.*

Relevant and timely information on critical and important aspects (the top 20 per cent of items on critical path) is needed to make appropriate decisions. As information is received and assessed, any performance issue should be immediately addressed as this will avoid further delays, costs, and reduce project risks.

With defined knowledge, it is possible to quickly identify key areas of concern, determine their root cause, provide possible solutions, and make the necessary decisions to rectify the situation. The administering parties (CC, PM, CAdmin) must ensure this occurs and make the necessary tactical decisions that will place the project back on track in alignment with the planned baseline. To do so, overall construction works' performance assessments must be conducted on a regular basis as these will determine how it is faring in terms of time, cost, quality, and output. It is of key concern to ensure that the critical path is maintained and works are not needlessly delayed with avoidable errors—non-conformances, incidents, or accidents.

Figure 5.23 outlines the 6D (decision-making) processes, commencing by gathering data from monitoring activities to processing of data and information and to making performance decisions, issuing CA and PA notices.

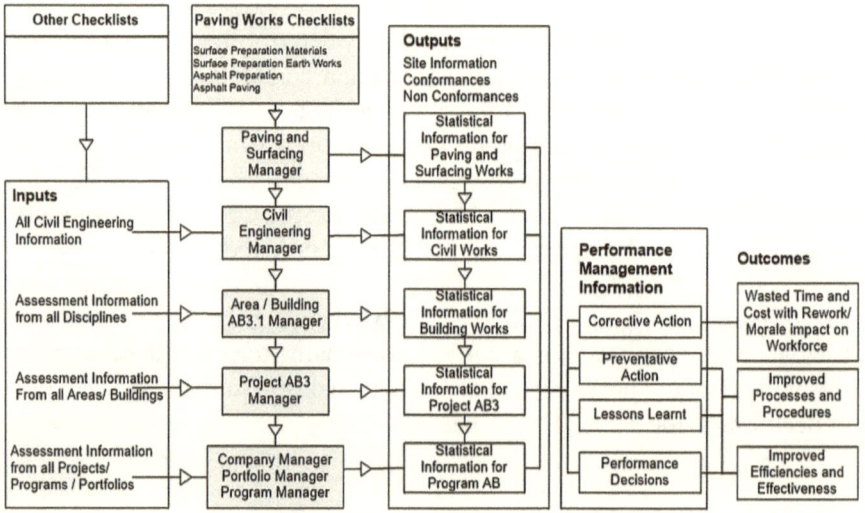

Figure 5.23 Process outlining how performance management information is derived.

In assessing progress, it is important to minimise delays wherever possible; however, it is also important to undertake a cost–benefit analysis with each decision to determine whether it is beneficial to fast-track or recover project time. Figure 5.24 is an example of a project schedule delay, where the actual progress is one month behind the baseline schedule. For the CC, the decision to recover and maintain the baseline should be based on cost comparison of accepting delays (penalty costs) with the additional costs that will be incurred (by adding resource, fast-tracking the works). For the CAdmin and CS, the decision to fast-track or expedite the works has additional risk implications, which if realised can impact the future work progress and possibly the ultimate viability of the project.

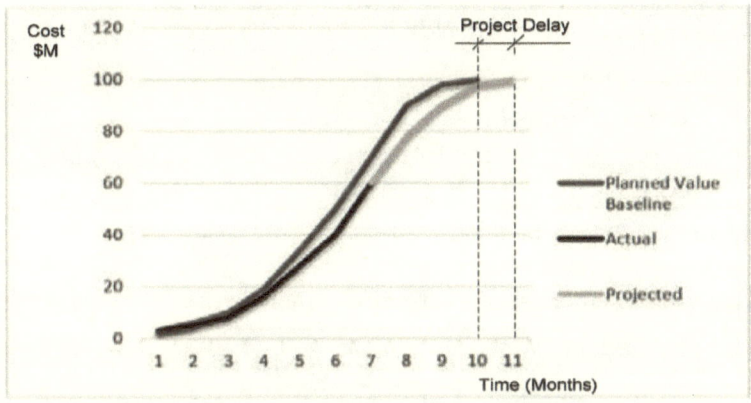

Figure 5.24 Example of S-curve graph showing actual project cost/time variance with baseline.

Of equal importance as to maintain the baseline/critical path is to ensure QSE compliance. There can be substantial cost implications for all parties involved in the contract if non-compliances are realised. However, unlike the time aspect, there is no applicable cost-benefit assessment to determine whether the non-compliance should be rectified. Non-compliances must be rectified at whatever cost by corrective or preventative action, which may involve additional works or rework (unless otherwise agreed by all parties).

The processing of key inputs, activities, and outputs of 6D (decision-making) are listed in the checklist below.

| 6D (Decision-Making) Checklist | ✓/✗ |
|---|---|
| **Key Inputs**<br><br>List key items required to assess performance, include the following:<br><br>• 4D and 5D reports for assessment period<br>  - STC (scope, time, and cost) aspects, with focus on critical path activities<br>  - QSE factors, with focus on non-compliance and required additional works or rework<br>  - 4M factors<br>  - notices and instructions issued (NC, CA, PA, and IOs)<br>  - other factors such as weather. | |

| 6D (Decision-Making) Checklist | ✓/✗ |
|---|---|
| • reference documents.<br>    - contract documents<br>    - list of priority of items based on client advice, programme (critical path), cost, or quality. | |
| **Performance Assessment Process**<br><br>Outline performance assessment process, include the following:<br><br>• Evaluate STC factors with respect to baseline and use and interpret earned value calculations (SPI, CPI), detailing the following:<br>    - compliant/non-compliant (SPI, CPI, rework, and rectification requirements)<br>    - progress (SPI, CPI, critical path, and schedule baseline impact)<br>    - other (additional works requested by client, issue resolution, approvals).<br><br>• Evaluate root causes and determine action using decision process tools.<br>    - flowcharts—process maps used to display the sequence of steps, activities, decision points, workflow branching and loops, the paths taken for particular outcome to occur<br>    - process decision programme charts (PDPC)—used to establish steps with contingencies for attaining certain goal<br>    - affinity diagrams—mind-mapping techniques used to generate ideas and form patterns of thought about a problem<br>    - matrix diagrams—used to perform data analysis within the structure by showing strength of relationships between factors, causes, and objectives<br>    - activity network diagrams—arrow diagrams used to determine schedule dependencies for particular works and establish critical path<br>    - prioritisation matrices—used to identify key issues, find suitable alternatives, and determine best option in terms of weighting priorities | |

| 6D (Decision-Making) Checklist | ✓/✗ |
|---|---|
| - relationship and interrelationship diagrams—used to map intertwined logical relationships for solving complex scenarios that possess for up to 50 relevant items, usually developed with data generated from affinity diagram, tree diagram, or fishbone diagram.<br><br>• Assess further risks, re-evaluate identified risks, and investigate newly arising risks from proposed measures to be undertaken<br>• Create forecasts and use milestones as control measures. | |
| **Measures to Improve Performance**<br><br>Outline improvement measures, include the following:<br><br>• Management factors.<br>   - measures to improve or reduce the quality aspects (noting that quality is a constraint which if reduced will enable works to proceed at a faster rate)<br>   - control processes (improve surveillance to reduce rework).<br><br>• Reschedule activities and reconfigure or re-plan the works and scope.<br>• Improve 4M factors.<br>   - manpower factors<br>      o Crash or compress schedule and fast-track the works.<br>      o Add resources, preferably with more experienced people, to undertake activities on the critical path.<br>      o Work extra hours with the same resources.<br>   - materials factors<br>      o Pre-order materials to avoid delivery delays.<br>   - machinery factors<br>      o Add more 'fit for purpose' equipment or plant to ease work operation.<br>      o Add plant including crane to facilitate the movement of site materials.<br>      o Increase power-to-weight ratio or load-carrying capacity of individual machines. | |

| 6D (Decision-Making) Checklist | ✓/✗ |
|---|---|
|     o  Use more appropriate tools, e.g. equipment and power tools with variety of subassemblies and interchangeable parts, smaller machines (for finishing operations), and all-purpose construction machines with increased reliability and longer service life.<br>- subcontractor performance factors<br>    o  As per manpower factors above, increase number of suitably experienced subcontractors and rate of work.<br>    o  Improve management of works.<br>    o  Reconfigure and compress work schedule.<br>    o  Fast-track activities and works.<br>    o  Overlap critical path activities where possible. | |
| **Outputs**<br><br>List key outputs, include the following:<br><br>- quantitative cost/benefit assessment (for cost-based decisions)<br>- notices and instructions issued (CA, PA, IO, CO)<br>- performance report aspects<br>    - CAdmin's monthly report to PM<br>        o  assess QSE aspects<br>        o  % efficiency, tables, histogram, pictures<br>        o  determine cost of NCs in terms of rework, supervision time, project delays<br>        o  assess efficiencies and effectiveness<br>        o  recommendations<br>        o  evaluations and conclusions from information.<br>    - PM's monthly report to the client<br>        o  list of decisions required to be made<br>        o  monthly performance overview in terms of Q-STC and HSE<br>        o  other project-specific issues to be resolved.<br>- updates to management plans and processes<br>    - Make updates to the project management plan and project documents.<br>    - Recalculate how much the project will cost to complete and how long it will take. | |

| 6D (Decision-Making) Checklist | ✓/✗ |
|---|---|
| - Manage the time and cost reserves and seek additional funds if required.<br>- Evaluate the effectiveness of risk responses in a risk audit.<br>• list of benefits of maintaining baseline performance.<br>   - tangible benefits—avoiding conflict with client and penalties, reducing cost of future business, reducing holding costs such as insurance cost and cost of retaining 4M to complete works (management and supervision costs)<br>   - intangible benefits—maintaining company image, improving possibility of repeat business, maintaining workforce morale, aligning company processes with the best practice. | |

Once poor performance is detected, all available information should be assessed to establish the relevant factors and possible options. Once determined, a cost/benefit assessment of each option should be considered in order to make the most appropriate cost-effective decision, including that of taking no action. Below is an example of cost benefit assessment in table form.

| Quantitative Assessment Example | | | |
|---|---|---|---|
| Cost/Benefit | $ (M) | Time (Weeks) | Notes |
| No Change, inclusive of Penalties/Delays | −5 | +4 | 4 weeks' delay, $5M penalties |
| 1. Improve QC Control | 1.5 | −2 | Required |
| 2. Reconfigure Schedule | 1.5 | −2 | Required |
| 3. Add 4M resources | 2 | −1 | Required |
| **Total Improvement with items 1, 2, and 3** | 5 | −1 | 5 weeks' saving, $5M cost |
| Contingency to completion | — | −1 | |

The above example summarises the costs and schedule variances of a particular assessment of available options from the CC's viewpoint, noting the cost of doing nothing, with no change equals the cost of meeting schedule ($5M).

# Chapter 6

# Phase 3: The Closing Phase of Construction

As construction works near completion, a project close-out process is required. This process should be pre-planned and detailed within the CS's management plan.

During the closing phase, the CS's role should remain focused on ensuring works undertaken are in accordance with the contract documents and are QSE compliant. The quality aspects should be checked prior to completing any activity, work package, or accepting these as completed. And during the works, the HSE aspects should be monitored for compliance. Even as commissioning works are undertaken, the CS must ensure that these works, including permanent and temporary equipment, materials, and consumable are QSE compliant.

The close-out phase is as important as any other phase, even more so, as often focus is lost and progress is slowed as allocated 4M resources are reassigned and demobilised to other works and projects. The key activities during the closing phase of construction are the following:

- Cx (commissioning) works, conducting performance tests, CxPI (Cx pre-inspection), start-up, FPT (functional performance test), and acceptance of operating systems

- defect-and-omission management, verifying completed works to accept work with minor defects and omissions, and management of defects and omissions during maintenance period
- CC's final submittals, as-builts (capturing technical modifications during construction phase), O&M (operation and maintenance) manuals
- completion and post-completion works, finalising contract claims, amounts, and quantities, and making completion payment.

Commissioning of installed equipment, plant, and equipment is a key step to finalising the construction works and making the construction project operational. Testing and commissioning may seem trivial and easy to accomplish; however, these can be overly complex and complicated by the applied technology that must be made to perform to specific criteria. How these are undertaken can determine the future viability of the built outcome as these influence the indoor air quality, health and safety of occupants (employee absenteeism and tenant turnover for commercial premise), and operating costs, which can be substantial if not undertaken properly.

## Cx (Commissioning) Works

Commissioning is a systematic process of ensuring that all building systems perform interactively according to the design intent and the owner's operational needs. This process begins at the design phase where the intent is documented and continues through construction, acceptance, and the warranty period, with actual verification of performance. The purpose of the Cx and performance testing is to:

- ensure the installed plant, equipment, and systems as specified, are installed, tested and made operational so these perform as intended, achieving contract requirements and, thereby, meeting client expectations
- demonstrate control systems are working in accordance with the current building operation plan, achieving required performance (HVAC, water, lighting), operating schedules, space environmental conditions
- ensure manufacturer procedures are followed so warrantees are not voided.

The following are the benefits of commissioning:

- achieves proper and efficient equipment operation
- enables coordination between design, construction, and operational requirements
- achieves optimal indoor air quality conditions, occupant comfort, and productivity, thereby also reducing the potential for liability relating to poor indoor air quality or HVAC problems
- reduces operation and maintenance costs.

Cx is therefore a crucial step in completing the construction works, as such, consideration should be given to the required processes and resources. Essential are the appointment of the CxA (commissioning agent) and the Cx plan.

## The CxA (Commissioning Agent)

Essential to undertaking Cx works is to have a suitably qualified, independent party who is able to verify the commissioning works are conducted as contracted. This role is to be undertaken by a CxA (commissioning agent), who undertakes a similar role to the CS but limited to overseeing Cx of equipment, plant, and systems. The primary role of the CxA is to assist in the development of and coordinating the execution of testing plan, observe, and document performance to ensure equipment and systems are functioning in accordance with the design intent and in accordance with the contract documents.

The complexity of the Cx often requires a concerted team effort that includes various parties as member of the Cx team. The Cx team should include the following:

- CC's representatives—respective subcontractor and suppliers, mechanical, electrical, plumber, piping, sheet metal contractors
- CxA—an independent party tasked with verifying commissioning works are conducted as contracted
- client's representative—CS, CAdmin, PM
- building operator—the plant operations or building mechanic, if known.

Ultimately, however, the Cx works are the responsibility of the CC, who must undertake the Cx works, provide all necessary tools, and conduct all tests. The CC must, therefore, provide skilled technicians who may also be made available to assist the CxA in completing the Cx programme. However, the time required for testing, the work schedule, is to be specified by CxA and coordinated by CC, who is to ensure skilled technicians are available, complete the necessary tests, make necessary adjustments, and rectify all problems and issues faced.

## The Cx Management Plan

As indicated in the previous chapters, it is important to plan all works, including Cx works. Like other plans, the Cx plan outlines the required process for the CC and serves as a management tool for the CS, CAdmin, and CS, who are to ensure the works are undertaken and completed as planned. The required Cx process can be derived from the contract documents (ABCD) and from CC's submittals, either during the bid stage or pre-Cx phase.

The CC is responsible for preparing the Cx plan and issuing it to the CS/CAdmin prior to the closing phases of construction works for verification and approval to use. The Cx plan may need to be updated and amended until it is sufficiently comprehensive, outlining all required processes as well as metrics to be used to measure quality. It may also need to evolve and expand until it is ready for implementation. Once implemented, it should be strictly adhered to by the contractor's Cx team.

Once the Cx plan is considered final, the CS/CAdmin must finalise the Cx management plan. Below is the checklist that can be used to review the Cx management plan.

| The Cx Management Plan Checklist | ✓/✗ |
|---|---|
| **Purpose of Cx Plan**<br><br>Outline the purpose of Cx plan, include the following:<br><br>• to ensure the intent of building owners and designers (as noted in contract documents) are incorporated as commissioning requirements<br>• to know in advance the scope and schedule of requirements of commissioning process | |

| The Cx Management Plan Checklist | ✓/✗ |
|---|---|
| • to establish Cx process and activities that will ensure the installed energy-related systems will function as intended, such as:<br>   - mechanical and passive systems<br>   - heating, ventilating, air conditioning, and refrigeration (HVAC&R) systems with associated controls<br>   - chilled-water system<br>   - lighting with daylighting controls<br>   - domestic hot-water systems<br>   - renewable energy systems (wind, solar).<br><br>• to ensure installed systems and equipment are working according to the specifications and in accordance with building operational requirements<br>• to outline responsibilities of each party, including outlining requirements for CxA (commissioning agent)<br>• to outline required meetings and reviews to ensure coordination with all parties, including equipment specialists from supply companies. | |
| **Scope of Cx Works**<br><br>Define scope of works and specific requirements by:<br><br>• conducting hazard and risk assessment of Cx activities, determining whether it is required to:<br>   - isolate areas during works<br>   - provide security and access control.<br><br>• appointing dedicated team to undertake Cx works, determining the following:<br>   - procedures required for proposed process and energy sources, such as gas, water, and electrical systems<br>   - procedures for handling failures at initial start-up or during commissioning.<br><br>• training requirements<br>   - for commissioning personnel<br>   - for end-user and maintenance personnel. | |

| The Cx Management Plan Checklist | ✓/✗ |
|---|---|
| **Key Inputs and Essential References**<br><br>List key inputs and essential references, include the following:<br><br>• relevant parts of the contract documents (ABCD)<br>    - basis of design, design intent, explanation of the ideas, concepts and criteria considered important<br>    - commissioning specification, operations, controls, and performance verification requirements.<br><br>• supplier control drawings or equipment documentation<br>• CC's submittals prior to Cx phase<br>    - Cx plan (with input from CxA), providing the following:<br>        o list of installed systems, assemblies, equipment, and components, listing also design changes that occurred during the construction phase<br>        o process and schedule for Cx works with detailed start-up procedures, test procedures, test protocols, and start-up checklists<br>        o SWMS for works identified as high risk<br>        o specific checklists and commissioning process test procedures for actual HVAC&R systems, assemblies, equipment, and components to be furnished and installed as part of the construction contract<br>    - testing plans detailing procedures and checklists for systems, subsystems, and equipment, providing the following:<br>        o calibration testing<br>        o testing for all operating modes, interlocks, control responses, and responses to abnormal or emergency conditions, verifying proper response of building automation system controllers and sensors<br>        o tests to be performed using design conditions whenever possible (the CC may require to simulated conditions using an artificial load, when it is not possible and practical to test under design conditions)<br>        o equipment to simulate loads, setting simulated conditions and specifying methods of simulation, including processes for the following: | |

| The Cx Management Plan Checklist | ✓/✗ |
|---|---|
| <ul><li>seasonal testing (after complete appropriate initial performance tests)</li><li>deficiency occurring outside scope of HVAC&R system.</li></ul><ul><li>shop drawing for equipment and plant to be commissioned.</li></ul><br>• CC's submittals during Cx phase<ul><li>certificates of readiness, certifying all referenced systems, subsystems, equipment, and associated controls are ready for testing</li><li>certificates of completion of installation, pre-start, and start-up activities</li><li>test records and as-built documentation<ul><li>Test data, inspection reports, and certificates with systems manual</li><li>Incorporate commissioning requirements into as-built documentation.</li></ul></li><li>corrective action rectification processes and documentation</li><li>completed commissioning documentation and reports, tests and inspection reports, and certificates.</li></ul> | |
| **Role and Responsibilities**<br><br>Outline roles and responsibilities for key appointments, detailing the following:<br><br>• The CS and CAdmin requirements during the commissioning phase include the following:<ul><li>prior to commissioning phase<ul><li>Review CC/CxA submittals.</li><li>Brief CC/CxA on works and installations.</li><li>Ensure all Cx activities are scheduled and issue notice of acceptance prior to the first scheduled system turnover.</li></ul></li><li>during the testing and Cx phase<ul><li>Ensure all measurement instruments are calibrated, including logging devices.</li></ul></li></ul> | |

| The Cx Management Plan Checklist | ✓/✗ |
|---|---|
| <ul><li>o Monitor all Cx works, including pre-Cx, Cx, functional testing, start-up and testing.</li><li>o Provide planning, scheduling, coordination support.</li><li>o Ensure all inspection, testing, adjustments, balancing of equipment and systems, and retesting are completed as required.</li><li>o Ensure any required additional Cx tests as directed by CxA are performed.</li><li>o Ensure all deficiencies are resolved, including items identified by CxA and as recorded on the CA register.</li><li>o Ensure Cx documentation is complete with CxA input included.</li><li>o Ensure data acquisition equipment is provided by CC to record data for the complete range of testing for the required test period.</li><li>o Ensure the safe operation of the equipment and systems from the time of initial energisation until acceptance of the overall project.</li><li>o Ensure that CC provides reports identifying all installed systems, assemblies, equipment, and components, including noting design changes that occurred during the construction phase.</li><li>o Ensure CC/CxA provides all required submittals, including systems manuals, completed checklists, such as manufacturer's pre-start and start-up checklists for systems, assemblies, equipment, and components to be verified and tested.</li></ul><ul><li>- post-commissioning</li><ul><li>o Review CxA's submittals, systems manuals, and other documents and reports.</li></ul></ul><ul><li>• CxA's role requirements are the following:</li><ul><li>- must have relevant experience of similar size, capacity, and complexity of plant and systems and must have skills to effectively complete the works</li><li>- be independent and able to verify that Cx works are conducted as contracted</li></ul></ul> | |

| The Cx Management Plan Checklist | ✓/✗ |
|---|---|
| - prior to commissioning phase<br>    o Review contractor submittals applicable to systems and equipment being commissioned.<br>    o Verify installation of systems and equipment, ensuring construction works are sufficiently completed so the systems can be started, tested, balanced, and commissioning procedures undertaken.<br>    o All equipment, materials, pipe, duct, wire, insulation, controls are as per the contract documents and related directives, clarifications, variations, and change orders.<br>- during the testing and commissioning phase<br>    o Direct Cx testing to ensure CC performs all required testing, including functional testing of all plant equipment, to confirm compliance with design basis and operating criteria.<br>    o Check installation, conduct preliminary mechanical and electrical checks, pressure-testing checks, flushing and cleaning of equipment, and piping check.<br>    o Check integrity of all connections and safety systems and verify that plant is functionally complete.<br>    o Ensure static tests are conducted to demonstrate interactive operation of components within the system prior to charging or energising these for the first time.<br>    o Inspect and test each component of large systems (such as chilled-water distribution network) prior to placing these into service.<br>    o Monitor start-up of all components, calibration of controls and equipment, tuning and initial operation of plant.<br>    o Verify that testing, adjustments, and balancing of work are complete.<br>- reporting requirements<br>    o Provide documentation of all checks, inspections, and test activities undertaken.<br>    o Report results and performance. | |

| The Cx Management Plan Checklist | ✓/✗ |
|---|---|
|     o  Issue systems manual for the commissioned systems.<br>    o  Issue commissioning report.<br>    o  Confirm that as-built drawings for system and plant are correct and complete.<br>- post-commissioning<br>    o  Review building operation within ten months after substantial completion.<br>    o  Verify staff training is conducted. | |
| **System and Equipment to Be Commissioned**<br><br>List all systems and equipment to be commissioned prior, include the following:<br><br>• mechanical equipment<br>    - piping<br>    - pumps<br>    - supply VAV (variable air volume) box<br>    - exhaust valve<br>    - control valves<br>    - supply air-handling units<br>    - exhaust fans<br>    - instrumentation<br>    - reheat coils (electric and hot water)<br>    - fan coil units<br>    - unit heaters<br>    - humidifiers<br>    - heat exchangers<br>    - field device controllers.<br><br>• mechanical and passive systems<br>    - chilled water<br>    - heating hot water<br>    - refrigeration<br>    - HVAC, including air-handling units, economisers, discharge air temperature reset, VAV modulation<br>    - control/building automation system and laboratory control system<br>    - space temperature, space pressurisation, building envelope | |

| The Cx Management Plan Checklist | ✓/✗ |
|---|---|
| - lighting systems, including all fixtures and their controls<br>- water System, including plumbing fixture and irrigation system dishwashers, cooling towers, evaporative cooling equipment, heating and cooling systems, humidification, and make-up water systems. | |
| **The Cx Staged Process**<br><br>Outline the scope of Cx works in stages, include the following:<br><br>• stage 1: planning<br>• stage 2: CxPI (Cx pre-inspection) works<br>• stage 3: equipment start-up<br>• stage 4: TAB (testing and balancing)<br>• stage 5: FPT (functional performance testing)<br>• stage 6: deficiency resolution<br>• stage 7: verification testing<br>• stage 8: certification. | |
| **Stage 1: Planning, Establishing and Testing Requirements**<br><br>Outline the requirements of the Cx process, as applicable, include the following:<br><br>• Cx team's key appointments and responsibilities for CxA, installer, testing, and balancing contractor, instrumentation/control specialist.<br>  - Establish authority to alter set points and direct sensor values when it is not practical to simulate conditions.<br>  - Detail what support is to be provided to Cx team—technicians, instrumentation, tools.<br><br>• Establish Cx and testing requirements.<br>  - Detail the type of measurements, whether short term or continuous, required to demonstrate compliance.<br>  - Use instantaneous measurements to demonstrate that systems, such as central heating and cooling systems, are sequencing and operate as intended.<br>  - Determine whether temporary data loggers or building automation system can be used to trend log data over time. | |

| The Cx Management Plan Checklist | ✓/✗ |
|---|---|
| - Set format of runtime reports to show load profile data, and demonstrate compliance with control loads and intended operating requirements.<br>- Detail the testing scope for the equipment installed, such as HVAC&R, testing to include installation, central equipment for heat generation and refrigeration, distribution systems to each conditioned space, with tests to measure capacities and effectiveness of operational and control functions. | |
| **Kick-Off Meeting Agenda**<br><br>A pre-Cx start meeting with the Cx team should be held to clarify the requirements of the Cx process. The meeting agenda should cover the following items:<br><br>• key appointments (CxA's team), roles, and responsibilities, outlining CxA's authority for witnessing testing and approvals<br>• reference documents, listing contractual requirements<br>• CC's commissioning plan, which is to include the following:<br>  - schedule and plan for commissioning installed plant and equipment<br>  - coordination with utility departments, agencies, and their consultants<br>  - submittals requirements—shop and working drawings, material samples<br>  - health, safety, and accident prevention measures<br>  - procedure to vary the works, where extra work and time extensions are required<br>  - lines of communication<br>  - material testing and acceptance procedures. | |
| **Stage 2: CxPI (Cx Pre-Inspection) Works**<br><br>After equipment is installed, connected, and ready to be operated but before balancing and FPT (functional performance testing), the CC should:<br><br>• confirm and validate that the installation of equipment for Cx is complete<br>• issue schedule of pre-inspection so CxA can witness parts or all of the pre-inspection | |

| The Cx Management Plan Checklist | ✓/✗ |
|---|---|
| <ul><li>ensure other trades, such as electrical, are present throughout the CxPI process to confirm connections and assist in the Cx process</li><li>conduct pre-Cx checking, testing, and adjusting building plant and engineering services so these meet specification, including visual inspection to confirm automatic controls, such as time clocks or direct digital controls</li><li>certify the results of the pre-inspection.</li></ul> | |
| **Stage 3: Equipment Start-Up**<br><br>For equipment start-up, ensure the equipment and systems are ready for commissioning, include the following:<br><br><ul><li>CC to issue comprehensive start-up programme prior to the first scheduled system turnover, with the following:<ul><li>detailed description of all equipment and systems to be tested and including operating parameters and control sequences</li><li>agreed action items from meetings with CxA</li><li>integrated start-up and construction schedule</li><li>scoped system drawings</li><li>turnover package list</li><li>master tracking system</li><li>start-up administrative manual</li><li>start-up technical manual, including comprehensive commissioning and test procedures</li><li>responsibility matrix</li><li>all functional performance test and checklists</li><li>all verification of tests and checklist</li><li>system/plant operating instructions</li><li>callout list with name and telephone numbers (including 24-hour emergency numbers)</li><li>Cx-CA (corrective action) procedure and forms</li><li>start-up process, which should be scheduled and executed as soon as possible after substantial completion of the system or subsystem but before balancing work is initiated</li></ul></li></ul> | |

| The Cx Management Plan Checklist | ✓/✗ |
|---|---|
| • CC to detail the start-up process with following steps:<br>  - Conduct pre-start inspection.<br>  - Equipment start-up (after pre-inspection).<br>  - Systematically start equipment up according to schedule.<br>  - Ensure representatives of equipment vendors are present to observe and to assist in the start-up process.<br>  - Start-up technician, the designated subcontractor, and the manufacturer's representative should be present at start-up and provide start-up report.<br>  - Report CA to be undertaken, and once resolved, certify work has been completed as required.<br>  - Test, balance system, and verify controls sequences after equipment has been successfully started up and operated.<br>  - Ensure all completed forms are reviewed by CxA prior to issue to CS. | |
| **Stage 4: TAB (Testing and Balancing)**<br><br>Outline the TAB procedure, include the following:<br><br>• Prior to testing, CC is to:<br>  - issue copies of reports, sample forms, checklists, and certificates (including calibration certificates)<br>  - issue proposed schedule for testing and balancing works and arrange CxA to witness testing and balancing work.<br>• During the testing and balancing, CC is to:<br>  - provide technicians, instrumentation, and tools to verify testing and balancing works as agreed by CxA<br>  - ensure the same instruments (model and serial number) are used as previously conducted tests<br>  - apply corrective measures as required.<br>    o For results with, say, a variance of more than 10 per cent or deviation by more than 3 dB, this shall result in rejection and require additional adjusting, balancing, and testing.<br>    o Inform CxA of all deficiencies remedied and those that are to be retested.<br>    o Conduct verification testing to ensure compliance. | |

| The Cx Management Plan Checklist | ✓/✗ |
|---|---|
| **Example of HVAC&R System TAB Procedure**<br><br>Outline HVAC&R TAB procedure and requirements, include the following:<br><br>• testing and acceptance procedures for equipment, instrumentation, and control system<br>• submittals, providing test data, record of inspection, and equipment certification.<br>• pipe testing procedure, detailing the following:<br>   - pipe system cleaning, flushing, hydrostatic tests, and chemical treatment process<br>   - sequence of testing and testing procedures for each section of pipe, identifying pipe by zone or sector identification marker (also to be noted on as-built drawings)<br>   - description of equipment required for flushing operations<br>   - minimum flushing velocity to be applied<br>   - tracking checklist for managing and ensuring all pipe sections are cleaned, flushed, hydrostatically tested, and chemically treated<br>   - report detailing works completed and test results.<br>• energy supply (steam, hot-water, and solar) systems' testing procedure<br>   - sequence and procedures for testing each equipment item and pipe section<br>   - list of required technicians, instrumentation, tools, and equipment to test performance.<br><br>• refrigeration (chillers, cooling towers, refrigerant compressors, and condensers, heat pumps) systems' testing procedure<br>   - sequence and procedures for testing of each equipment item and pipe section<br>   - list of required technicians, instrumentation, tools, and equipment to test performance. | |

| The Cx Management Plan Checklist | ✓/✗ |
|---|---|
| <ul><li>HVAC&R distribution system testing procedure<ul><li>list of required technicians, instrumentation, tools, and equipment to test performance of air, steam, hydronic distribution systems, special exhaust, and distribution systems, such as HVAC&R terminal equipment and unitary equipment.</li></ul></li><li>vibration and sound testing procedure<ul><li>list of required technicians, instrumentation, tools, and equipment to test performance of vibration isolation and seismic controls.</li></ul></li></ul> | |
| **Stage 5: FPT (Functional Performance Testing)**<br><br>FPT is to be conducted to validate component and systems performance and to allow deficiencies to be corrected prior to handover. The FPT are scheduled after systems are complete and TAB is complete. The scope of FPT includes the following:<br><ul><li>verification tests to ensure plant, systems, and services are operating in accordance with the design intent</li><li>participation by CxA to confirm procedure, data sheets, witnessing of tests, and certify results</li><li>recording all measurements and results and providing a comprehensive summary describing the operation at the time of testing</li><li>additional commissioning activities depending on results, which may be required after system adjustments and replacements</li><li>use of measuring instruments and logging devices to record test data over agreed period, testing continuously over a given period.</li></ul> | |
| **Stage 6: Deficiency Resolution**<br><br>The deficiency resolution process should include the following:<br><ul><li>detailed additional work to be undertaken by CC with input from equipment supplier and CxA, such as:<ul><li>testing performance under varying loads</li><li>rectifying issues such as misalignment of equipment installation and interface issues</li><li>undertaking additional Cx work</li></ul></li></ul> | |

| The Cx Management Plan Checklist | ✓/✗ |
|---|---|
| <ul><li>timing for completion of corrective work, with possible reschedule of completion of Cx process</li><li>CxA is to control corrective works scope, noting the following:<ul><li>Approval should be obtained for experimentation.</li><li>CxA is to determine nature of the problem, outline steps to be taken, and suggest timings for completion of activities.</li></ul></li></ul> | |
| **Stage 7: Verification Testing**<br><br>Verification testing is conducted to verify components, equipment, systems, subsystems, and interfaces between systems to ensure these are operating in accordance with contract requirements. Testing of all operating modes, interlocks, specified control responses, specific responses to abnormal or emergency conditions and verification of the proper response of the building automation system controllers and sensors are to be undertaken. Generally, the CxA outlines the requirements for the verification testing scope and schedule, which include the following:<br><br><ul><li>Roles and responsibilities for Cx team members, manufacturers, suppliers, CC, mechanical subcontractors, technicians, ensuring these are familiar with the construction and operation of system.</li><li>Documentation and reporting.<ul><li>providing checklists for each component, piece of equipment, system, and subsystem, including all interfaces, interlocks</li><li>recording requirements for the following:<ul><li>results for each tested item (each should be on different entry line with space provided for comments)</li><li>modes of operation (each with separate checklist, noting whether mode under test responded as required and confirmed by all necessary parties).</li></ul></li><li>submittals, test procedures, and data sheets for review by system designer.</li></ul></li></ul> | |

| The Cx Management Plan Checklist | ✓/✗ |
|---|---|
| <ul><li>Instrumentation.<ul><li>calibration of measurement instrumentation</li><li>ensuring tests are undertaken within manufacturer's recommended period for testing.</li></ul></li><li>Verification procedures and systems.<ul><li>witnessing and verification of operating tests and checks for equipment</li><li>ensuring completed operating cycle is tested, on normal shutdown, normal auto-position, normal manual position, unoccupied cycle, emergency power, and with alarm conditions</li><li>ensuring operating checks include safety cut-outs, alarms, and interlocks with smoke control, refrigeration monitoring system, and life-safety systems tested for all modes of operation of the mechanical system</li><li>inspection and verification of position of each device and interlock, ensuring each is signed as acceptable (yes) or failed (no)</li><li>providing appropriate comments to checklist if operating deficiency is observed</li><li>for each controller or sensor, ensuring monitoring and control system reading and test instrument reading are recorded</li><li>for readings outside the control range, detailing actions to be taken to remedy the deficiency and to ensure equipment is checked, adjusted, recalibrated as required, and recording of results on checklist prior to retesting</li><li>the range in which results should be regarded as failure and rejection of TAB (testing and balancing) activity</li><li>requirement for CxA to witness field verification of the final testing, adjusting, and balancing.</li></ul></li><li>The TAB final report should only be validated after CxA verifies operating system is functioning in accordance with the contract documents.</li></ul> | |

| The Cx Management Plan Checklist | ✓/✗ |
|---|---|
| **Stage 8: Certification**<br><br>Outline the CC certification requirements for the installed equipment and systems, include the following:<br><br>• testing certification<br>　- set systems, subsystems, and equipment in all operating modes, on normal shutdown, normal auto-position, normal manual position, unoccupied cycle, emergency power, and with alarm conditions<br>　- safety cut-outs, alarms, and interlocks with smoke control and life-safety systems for each mode of operation<br>　- testing, adjusting, and balancing procedures completed, discrepancies corrected, and corrective work approved.<br><br>• equipment and system certification<br>　- installed systems, subsystems, equipment, instrumentation, and controls, confirming these were calibrated, started, pre-tested at set points established and are operating according to the contract documents<br>　- compliance certifications, to include all test results and detail all adjustments and balancing works. | |
| **Outputs**<br><br>The CC is to provide the following:<br><br>• test and inspection reports and certificates, include:<br>　- verification of testing, adjusting, and balancing reports<br>　- completed checklists of CxPI (commissioning pre-inspection) and FPT (functional performance test) with results<br>　- CA (corrective action) procedures and action taken to rectify deficiencies<br>　- certification of works<br>　　　o certificate of readiness, certifying that HVAC&R systems, subsystems, equipment, and associated controls are ready for testing | |

| The Cx Management Plan Checklist | ✓/✗ |
|---|---|
|     o  certificate of completion of all CA (corrective action) items<br>    o  certificate of completion, certifying installation, pre-start checks, and start-up procedures are complete and functioning as specified<br>    o  certificate of completion of installation and completion of commissioning activities<br><br>• reports and drawings, include:<br>    -  as-built drawings, coordination drawings, final instrumentation, and controls installation drawings<br>    -  O&M (operation and maintenance) manuals for building plant and systems<br>    -  operator training manuals<br>    -  valve tags and charts<br>    -  list of spare parts provided and location stored within building.<br>• guarantees and warranties, ensuring guarantees and warranties are not commenced until final acceptance of the commissioning of the major systems, auxiliary support systems, building system, such as HVAC, electrical, life-safety system, and all other systems included in the commissioning plan. | |

Where there are discrepancies between the actual outcome (what is occurring on-site) and the baseline, then the plan should be amended and updated accordingly to ensure the required performance of the equipment, plant, and systems are achieved.

## Cx Checklists and Notices

As the Cx works proceed, then the CxA should be directing the Cx for all items, equipment, components, and systems as per the Cx Plan. Of importance is the CxPI (Cx pre-inspection) assessment, followed by the start-up and FPT (functional performance testing) and verification testing. Below is an example checklist for CxPI for Hydronic Pump and FPT for AHU.

| CxPI (Commissioning Pre-Inspection)/FPT (Functional Performance Test) Example | ✓/✗ |
|---|---|
| **CxPI Example Checklist for Hydronic Pump**<br><br>Conduct CxPI by verifying installation is complete prior to operating, confirming the following:<br><br>• installation complete and verified<br>   - as per manufacturer's recommendations, in accordance with contract documents and approved shop drawings<br>   - pump and motor mountings checked<br>   - alignment visually checked<br>   - seismic vibration isolation pads installed<br>   - inertia base floating (with no vibrations).<br><br>• labelling (all equipment is clearly labelled)<br>• safety check complete<br>   - equipment guards are installed<br>   - disconnect switch installed<br>   - adequate access and clearance for maintenance provided.<br><br>• operational settings check<br>   - local valving set for normal operation<br>   - pumps equipped with isolation valves<br>   - pressure gauges installed in supply and return piping<br>   - provision for motor overload protection provided<br>   - discharge check valve installed<br>   - piping configuration as per detail<br>   - pump housing equipped with air bleed valve.<br><br>• control check, controls connection complete<br>• special requirement—pump housing insulated (where applicable). | |

| CxPI (Commissioning Pre-Inspection)/FPT (Functional Performance Test) Example | ✓/✗ |
|---|---|
| **FPT Example Checklist for Rooftop Air-Handling Unit**<br><br>Conduct FPT as per commissioning plan and contract documents, include the following:<br><br>• operations check<br>   - basic function verification<br>   - correct fan rotation<br>   - operational sequence verification<br>   - noise (ensure no excessive noise or vibration).<br><br>• sequence check<br>   - AHU offline sequence<br>   - status indication (filters, fans)<br>   - safety<br>   - alarms<br>   - start/stop<br>   - volume control<br>   - supply air temp control<br>   - ventilation setback<br>   - occupied mode<br>   - unoccupied mode<br>   - economiser mode<br>   - morning warm-up<br>   - operation on standby power.<br><br>• output check (confirm output results)<br>   - cooling<br>   - heating<br>   - air volume. | |

### Cx-CA (Corrective Action) Notice

As the Cx works proceeds, the CxA is to identify and prepare list of items, components, and systems requiring CA (corrective action). These items can include additional work required during start-up, FPT, and verification testing. The CC is then responsible for taking immediate CA measures to rectify deficiencies to achieve compliance with the contract documents and CA notices. Once completed, the CC is to notify the CxA, who should then recheck and re-verify that CA items have been corrected satisfactorily.

| Cx-CA Sheet | | | | | | |
|---|---|---|---|---|---|---|
| System or Turnover Package | | | | | | Sheet No. |
| Item Identifier | | | | | | |
| CxA Name | | | | | | Date |
| Equip. No. | Description for Condition | Recommended Corrective Measure | Action By | Action Due Date | Date Completed | Name and Signature of Person Completed Item |
| | | | | | | |
| | | | | | | |

Once the commission works are finalised and the majority of construction works completed, then the project can be assessed as 'operational ready'. Prior to accepting the completed works, it is important to determine whether there are major or minor defects and omissions and whether these are acceptable to the client for handover of the project to occur.

## Defect-and-Omission Management

Once the works are sufficiently completed, with practical completion certificate issued, and the works are handed over to the client, then the maintenance period commences; the defect and omission rectification works begins. These works are not additional or reworks, but rather rectification and minor repair works to achieve compliance. These works are necessary when CC has not achieved a 'defect free' completion as component of the works were assessed as not meeting the specifications. As with CA and PA notices, defects noted are not changes to the contract or scope of works; these are part of the works that must be completed prior to the issue of final certificate. Below is the checklist that can be used to review and manage defect and omission works.

| Defect-and-Omission Management Checklist | ✓/✗ |
|---|---|
| **Confirm Requirements**<br><br>Review requirements, noting the following:<br><br>• There is no contractual requirement for CS to list omissions and defects, but this should be done to ensure omissions and defects are catalogued and rectification work is properly monitored.<br>• The defect list prepared should be reviewed progressively, with each review issued to CC.<br>• The first inspection should be scheduled as soon as CC gives notice of the anticipated date of completion. | |
| **Manage Safety during Defect Rectification Works**<br><br>Invariably, defects occur after building becomes operational; therefore, the CC is to:<br><br>• conduct additional hazard/risk assessment and establish procedures for completing defective works in operational areas<br>• provide SWMS with specific safety systems to be employed, considering that rectification works will be undertaken in operational areas, include the following:<br>   - details of how works are to be undertaken<br>   - personnel to be utilised to rectify the defects<br>   - whether defect rectification crew will be permanently on-site or attend on an ad hoc basis<br>   - training requirements for defect rectification crew<br>   - access requirements for defect rectification crew. | |
| **Items Required for Handover and Operations**<br><br>List key items required to be completed prior to handover, include the following:<br><br>• Works are assessed by client as substantially complete and ready for operations.<br>• Connection to utilities and services is completed.<br>   - Water is supplied and taps are operational.<br>   - Electrical RCD and safety cut-out are installed.<br>   - Waste management system is operational (sanitary fittings). | |

| Defect-and-Omission Management Checklist | ✓/✗ |
|---|---|
| • The following certification of works is issued.<br>   - statutory certificate of occupancy/completion<br>   - certificate for self-certification of the works, certificate of compliance (as applicable)<br>   - electricity certificate<br>   - survey certificate<br>   - termite treatment certificate<br>   - road clearance provided<br>   - plumbing compliance certificate<br>   - gas plumbing certificate.<br>• Security is ready for transfer; keys are provided and are tagged.<br>• Warrantees and guarantees are received. | |

The CS should record all defects and omissions and track all items identified as not in compliance and requiring the attention by CC. All items should be listed in the register showing current status of defects, which should be regularly updated as defects are rectified or as new defects become apparent during the maintenance period, with each new version issued to the CC each time. Below is the register template sample.

| Defect-and-Omission List Register (Template) | | | | | |
|---|---|---|---|---|---|
| ID/ Location | Observation/ Action Required | Date to Be Completed by Contractor | Action Proposed and Taken by Contractor | Status<br>R = Rectified Date<br>O = Ongoing<br>P = Pending Solution | Comments |
| | | | | | |
| | | | | | |

The defect-and-omission register should note in the comments column any special access arrangement required to complete the works. This should be discussed with the client and occupiers of the premises prior to issue. The register should also include all CC's required submittals prior to issuing of final certificate, such as as-builts and O&M manuals.

## CC Final Submittals

Prior to issue of final certificate, the CC is required to provide final submittals—as-builts and O&M manuals. These submittals must then be reviewed by CS and CAdmin to ensure these are in approved format, including all required information, and are accurate and correct. Typical areas requiring review and confirmation for correctness include the following:

- dimensions and levels
- construction details
- product and equipment references
- references to other drawings and standards
- incorporation of approved design changes.

Where final submittals are found to be inadequate or inaccurate, these should be returned to CC for correction and resubmission.

## As-Built Drawings

As-built drawings should be prepared as works are undertaken, with service lines and items that are later hidden by other works, progressively marked up as works proceeds, and finalised at completion after final system checkout. Generally, as-builts are the updated shop drawings, showing exact installation. The installed system should therefore not be considered complete until the as-built information is received, reviewed, and approved. Below is the as-built drawing completion template sample.

| As-Built Drawing Completion Certificate (Template) | |
|---|---|
| Certification of As-built Drawing  The drawings listed below have been reviewed with the construction site works and are certified as inclusive of all site changes and verified as correct and complete. | |
| List of Drawings (drawing number, title, revision number and date, details | |
| Construction Supervisor (Signature and Date) | |

## O&M (Operation and Maintenance) Manuals

The O&M manuals should be provided for all equipment and system commissioned, including the following:

- systems (chillers, water system, condenser water system, heating system, supply air systems, and exhaust systems)
- equipment (pumps, chillers, boilers, control valves, expansion tanks, coils, and service valves).

The specific requirements for O&M manuals (content and format) are usually detailed in the contract documents. Below is a content checklist for reference.

| O&M Manuals Contents Checklist | ✓/✗ |
|---|---|
| **The O&M Content** <br><br> The O&M Manuals should include the following: <br><br> • description of design intent and operational policy <br> • final commissioning report and documents, in the following format: <br>    - system type 1 (chiller system, packaged unit, boiler system, etc.) <br>       o design narrative and criteria, sequences, approvals for equipment <br>       o start-up plan and report, approvals, corrections <br>       o pre-functional checklists <br>       o equipment types (fans, pumps, chiller, etc.) <br>       o functional tests (completed), trending and analysis, approvals and corrections, training plan, record and approvals, functional test forms, and a recommended recommissioning schedule. <br>    - system type 2 (repeat as per System type 1) <br><br> • manufacturer's operation, service, and maintenance manuals <br> • manufacturer's spare parts list and ordering procedure <br> • operational and maintenance routines, including copies of charts, procedures posted in plant rooms and building <br> • line diagrams of plant control systems as-built <br> • schematic layouts locating items such as valves and switches <br> • schedule of installed equipment <br> • list of as-builts, record drawings, or schematics for each item of plant or equipment, listing also safety features and the like | |

| O&M Manuals Contents Checklist | ✓/✗ |
|---|---|
| • emergency procedure, names, and contact details of maintenance personnel on call<br>• a list of tools, keys, and any special requirements<br>• insurance, guarantees, test certificates, and reports<br>• inspecting authority's certificates and reports. | |

Prior to completion, the O&M manual should be reviewed to ensure the following:

- There is compliance with requirements, noting all deficiencies for correction.
- Validity of warranty for all equipment is verified.

Once the O&M manuals are confirmed as suitable, all equipment commissioning data, together with the Cx report, should be placed into the respective section within the O&M manual.

## Completion and Post-Completion Activities

The end of the construction works marks the completion phase and commencement of the maintenance period or defects liability period. For issue of practical completion certificate, the works must be substantially complete and, if possible, defect-free. Issue of this certificate formally signifies that construction works are complete and the facility is fit and safe to occupy and use.

Prior to issue of certificate and prior to client taking possession, operational commissioning should be conducted. Operational commissioning is the readying of a new facility for its intended use, which requires operating all systems as intended prior to occupation and continued until facility is operational.

Upon issue of certificate, ownership and responsibility for the works (such as security, insurance, safety, and energy costs) then passes to the client. It is therefore essential that this be first coordinated with the client in order for the client to have sufficient time to make suitable arrangements for insurance of works and be in position to accept handover of the works.

During this post-completion, the maintenance period, which is normally 12 months, the CC should remain responsible for remedying all identified defects that occur within the building and its systems. At completion of the works and issue of practical completion certificate, the CS should be required to provide a statement as follows:

| Final Inspection Certificate (Template) | |
|---|---|
| I certify that [CS name] of [Company name] has carried out the final inspection concerning Contract [Name and Number] and that the Defects and Omissions listed on Register No. ### dated ### have been satisfactorily rectified and completed. | |
| Construction Supervisor (Signature and Date) | |

In the instances where the CC is to self-certify the work (where tasked also with ensuring QC of the works), the CC should also be required to provide certificate of self-certification. The certificate should be based on the template below.

| Certificate for Self-Certification of the Works (Template) | | | | |
|---|---|---|---|---|
| In accordance with the Conditions of Agreement, [CC name] of [Company name] certifies that the works are substantially complete and subject to the making good of any outstanding items detailed below or to any defects described under the Conditions of Agreement of [Contract No. and Title] which arise during the maintenance period. | | | | |
| List of exceptions | | | | |
| Name, Company Date, Company Seal, Signature, Date: | | | | |
| Item No/ Location | Defect item, Omissions, or Outstanding items | Proposed Action to be Taken by Contractor | Planned Date for Completion by Contractor | Actual Completion Date/ Comments |
| | | | | |
| | | | | |

Similarly, a certificate of compliance should be requested from CC when it is either within the contract scope to provide or when the CC has failed to provide the required certificates from third parties for the works. The certificate should be based on the template below.

| Certificate of Compliance (Template) | |
|---|---|
| Certificate of Compliance<br>The works undertaken are compliant with the construction documents and statutory requirements and are verified as complete without error or omission, with exception of items listed. | |
| List of exceptions (items requiring rectification) | |
| Name, Company Date, Company Seal, Signature, Date: | |

For large projects, with phased completion, partial handover may be required. In such instances, the client may take possession of parts of the constructed works, where these are assessed as fit for use and occupation, whilst the remaining areas still under construction are in CC's control.

As construction works near completion, prior to handover, a completion and post-completion plan should be devised and enacted. This plan should list key activities and procedures required to close the project, including rectifying defects and undertaking post-occupancy adjustments as required. The completion and post-completion key activities are listed in the checklist below.

| Completion and Post-Completion Checklist | ✓/✗ |
|---|---|
| **Practical Completion**<br><br>Review contract documents and list all items required to be finalised for issue of practical completion certificate, include the following:<br><br>• Receive all certificates and approvals.<br>    - statutory approvals, permits, certificate to occupy, completion certificate<br>        o DA (disability and access) compliance certificate<br>        o food safety<br>        o occupancy certificate<br>        o electrical/fire alarm/fire sprinkler certificates.<br>    - non-statutory approvals, certification of the works, certificate of compliance<br>    - manufacturers warranties<br>    - contractor's guarantees and warranties. | |

| Completion and Post-Completion Checklist | ✓/✗ |
|---|---|
| • Finalise all compliance tests.<br>    - pressure tests<br>    - equipment tests<br>    - sanitisation reports<br><br>• Complete Cx of all equipment and systems.<br>• Complete on-site training of maintenance personnel for use of equipment and systems.<br>• Receive CC's final submittals.<br>    - O&M manuals<br>    - as-built documents<br>    - Cx reports, certificates, and documentation.<br><br>• Finalise all outstanding matters with CC, ensuring the following:<br>    - all COs complete and executed<br>    - all claims resolved and finalised.<br>    -<br><br>• Conduct handover inspection with CS, CC, and client to:<br>    - ensure all required CA items, defects, and omissions are rectified and completed, and remainder are scheduled for rectification during maintenance period<br>    - obtain client acceptance for facility and for issue of handover certificate.<br><br>• Obtain insurance underwriter's approval for client to commence insurance coverage prior to handover.<br>• Prior to commencement of maintenance period.<br>    - Obtain schedule for rectification of remaining defects and omissions.<br>    - Obtain contact numbers for services trades.<br>    - During maintenance period, a 24-hour on-call service phone number for urgent works is required.<br><br>• Finalise contract amounts and quantities to make completion payment and release security as per contract (usually 50 per cent). | |
| **Project Close Out/Final Certificate**<br><br>Prior to closing the project and issue of final certificate, finalise all contractual matters, include the following: | |

| **Completion and Post-Completion Checklist** | ✓/✗ |
|---|---|
| <ul><li>Conduct final inspection and confirm all previously identified deficiencies completed.</li><li>Receive outstanding compliance certificate and inspections reports.</li><li>Issue completion certificate.</li><li>Release final security as per contract (usually 50 per cent).</li><li>Rate and report CC performance, assessing the following:<br>- technical and management capability and experience (in terms of 4M items)<br>- actual performance against initial baseline<br>- quality, number of issued notices (CA, PA, IO), noting areas where performance issues arose and needed attention<br>- HSE performance<br>   o number of incidents, accidents, near misses (Were these reported in a timely manner?)<br>   o Were safety meetings held on a regular basis?<br>   o Were safety audits and inspections conducted as scheduled?<br>   o Were the safety factors of commissioning conducted appropriately?<br>- performance during defects liability/maintenance period<br>- ability to undertake similar and suitability to prequalify for larger works.</li><li>Review design and contract documents to determine areas for improvement and document for future reference, considering the following:<br>- number of unidentified HSE risks and issues that arose as a result of design or documentation during the construction and commissioning works and once occupied, or in operation<br>- number of CC claims and contractual issues that arose as a result of error in documentation.</li></ul> | |

Whether the CC has achieved the required contractual completion is a matter for CAdmin/PM to evaluate in consultation with the client and to finalise with issue of final payment certificate.

# INDEX

## A

accidents  52, 58, 64, 69-70, 85, 99, 102, 109, 111, 120-1, 127, 148, 185
agreement  11, 19, 23, 29, 42, 46, 103, 182
air quality, indoor  155-6
alarms  171-2, 175, 183
as-builts  11, 22, 67-8, 72, 75, 155, 163, 168, 173, 178-80
audits  12, 14, 37-8, 68, 73-4, 78, 83-4, 98, 101, 106, 109, 111

## B

baseline  18, 38, 43, 45, 61, 67, 72-4, 99, 102, 111, 113, 126, 129, 145-6, 148-50
bill of quantities  11, 19, 42
building surveyor  12, 30-1

## C

CA (corrective action)  12, 24, 29, 38-9, 45, 67, 69-70, 74-5, 78, 98-100, 104-10, 112, 122-3, 172-3, 175-6
CAdmin (contract administrator)  13-14, 21-3, 25-6, 32-4, 46-9, 65, 67, 69-70, 76, 96, 105, 114, 116, 147-8, 156-7
calibration  41, 80, 82, 159, 162, 171
CC (construction contractor)  21-2, 24-6, 28-30, 34-5, 48-9, 66-8, 75-6, 81, 95-9, 114-18, 156-7, 161-2, 175-9, 182, 184-5
checklists  10, 22, 50, 53, 65, 94, 117-18, 127-8, 130, 135, 145, 149, 159, 166-7, 170-1
clients  18-21, 25-6, 49, 61, 66, 75, 80, 96, 98-9, 103, 150, 152-3, 176-8, 181, 183-5
CO (change order)  67, 107, 113, 135, 152
commissioning phase  41, 160, 162
commissioning plan  12, 68, 165, 173, 175
communications  15-17, 19, 23, 26, 32, 44-8, 95, 97, 130, 146, 165
completion  12, 16, 18, 22, 31, 38, 68, 75, 97, 105, 153-5, 160, 170, 177-9, 181-3
conditions  19, 24, 34, 77, 91, 93, 176
conformance  13-14, 16, 66-8
construction phase  31, 48, 63, 67, 70-1, 79, 87, 89, 96, 115-16, 155, 159, 161

Construction Specification  11, 19, 29, 42-3
contract  11-14, 19, 21-6, 34-5, 39-40, 42-3, 63-9, 76-7, 89-90, 96-8, 104-5, 113-16, 156-7, 174-6, 184-5
contractor  10, 12, 20, 41, 48, 79, 82, 90, 112, 126, 147, 178, 182-3
corrections  179-81
costs  11-12, 15-21, 29, 32-5, 42, 44, 61-2, 64-5, 88, 113, 125-6, 136-7, 147, 149-50, 152-3
CPI (cost performance index)  12, 126, 150
CS (construction supervisor)  13-14, 21-6, 28-9, 31-5, 47-9, 53, 59-60, 64-7, 69-70, 80, 96, 101-2, 120-1, 156-7, 177-9
CxA (commissioning agency/agent)  13-14, 32-3, 68, 75, 156-61, 164-7, 169-71, 173, 175-6
CxPI (commissioning pre-inspection)  154, 164-5, 172-5

## D

D cycle  11, 45, 59-61, 103, 124-9, 134, 144, 147-9
data  9-11, 14, 18, 22, 45, 59-62, 70, 79-80, 116, 124-31, 147-8, 150-1, 160-1, 164-5, 168-70
data gathering  11, 61, 70, 87, 126-8, 148
data processing  11, 61, 70, 126, 128-9, 148
decision-making  10-11, 39, 59-61, 103, 124, 126, 129, 144, 147-9
decisions  18, 25-6, 34, 44-5, 61, 63, 95, 111, 125, 127, 134, 147-8, 152
defects  15, 18, 39, 64, 68, 75, 88, 95, 136, 155, 176-8, 181-5

deficiencies  49, 69, 106, 144, 147, 160-1, 167, 169, 171-2, 176, 181
delays  10, 19, 24, 43, 58, 60, 147-8, 152-3
Deming, W. Edward  12, 17
design  21-2, 25, 55, 65, 72, 98, 106, 129-30, 156, 159, 180, 185
documentation  19, 52, 66-7, 99, 102, 106, 127, 160, 162, 170, 184-5
Drawings  11-12, 14, 19, 22-3, 29-31, 37, 41, 67-8, 71-2, 94, 106, 116-18, 165-6, 173-4, 179-80

## E

employees  27, 54, 56, 69-71, 77, 91, 112
equipment  13, 15, 30-1, 41, 48, 75, 82-3, 85, 91-3, 117-20, 155-6, 158-63, 165-74, 179-81, 184
errors  10, 15, 23, 88, 148, 183, 185
excavations  28, 52, 80, 92, 118
experience  16, 22, 35, 41, 54, 56-7, 59-60, 70, 95, 111, 121, 185

## F

failure  21, 49-50, 57, 109-10, 116, 124, 136, 146, 171
FPT (function performance test)  154, 164-5, 169, 172-5

## H

handover  48, 64, 75, 169, 176-7, 181, 183-4
hazards  13, 28-9, 53-6, 69, 73, 83, 101, 119, 158
HSE (health, safety, and environment)  12, 19, 33, 41, 49, 52, 58, 61, 88, 94, 107, 111, 122, 128-9, 136

## I

incidents  25, 50, 52, 64, 69-70, 76, 86, 90, 101-2, 106-7, 120-2, 127, 136, 148, 185
inspections  14, 24-5, 27, 34, 38, 41, 74, 76, 93, 101, 106-7, 114, 116, 161-2, 185
installation  29, 50, 57, 118, 160, 162, 165, 173-4
Integration  15-17, 19, 32
ITP (inspection and test plan)  13, 16, 24, 30, 36, 38, 67-8, 71, 73, 78-9, 82-3, 90, 103, 116, 127-8

## K

knowledge  8, 11, 15-20, 22, 29, 32-3, 59-62, 64, 66, 88, 124, 126, 144-5, 147

## M

machinery  11, 24, 30, 44, 47, 50, 57, 72, 74, 80, 93, 95, 98-9, 102, 120
manpower  11, 24, 30, 44, 47, 50, 56, 72, 74, 79, 91, 95, 98-9, 102, 119
materials  11, 23-4, 30, 38, 41-4, 47, 50, 57, 68, 71-2, 80-1, 92, 95, 98-100, 102-3
Me'Awel  11, 44
meetings  23, 46, 64, 71, 90, 95-9, 102-3, 166, 176

## N

NC (non-conformances)  24, 38-9, 45, 67-8, 70, 75, 78, 99-100, 102, 104-8, 110-11, 115, 127, 135-6, 148-9

## O

O&M (operation and maintenance)  14, 67-8, 155, 173, 178-81, 184
O&M manuals  67, 178-81, 184
omissions  18, 64, 68, 70-1, 75, 88, 126, 155, 176-8, 182-4
outputs  15-16, 41, 52, 58, 66-8, 88, 125, 127-30, 135-6, 145-7, 149, 152, 172, 175

## P

PA (preventative action)  24, 38-9, 41, 45, 67, 70, 75, 78, 84, 99-100, 102, 105, 107-12, 145, 148-9
PDPC (process decision programme charts)  150
performance  15, 19, 46, 62, 66-7, 87-8, 112, 116, 123, 125-7, 145-6, 149, 153, 155, 185
PM (project manager)  14-15, 21-2, 32-4, 46-9, 65-7, 69-70, 76, 96-7, 99, 103-5, 107, 121, 147, 152, 156
PMI (Project Management Institute)  14-17, 32
PPE (personal protection equipment)  27, 30, 51, 72, 79, 91, 93
precision  38
prevention  15, 22, 49, 52, 78-9, 84, 91, 120, 165
Procurement  15-17, 19, 32
project  9, 14-15, 17-18, 20, 31, 35, 42-4, 116-17, 123-4, 130, 136, 143-4, 147-8, 154, 176
project knowledge area  15, 32
project management  15, 17, 64-5, 152
proposal  26, 28-9
protection  30, 57, 72-3, 85

## Q

QA (quality assurance)  16, 34, 36, 41, 82, 112, 116
QC (quality control)  16, 26, 34, 38, 40-1, 58, 70-3, 79, 82, 94, 98-9, 128-9, 135-6, 153, 182
quality  12, 15-20, 23, 29, 32-5, 37-9, 42, 44, 61-2, 67, 74, 77, 87-8, 100-1, 150-1

## R

registers, site  27-8, 90
responsibilities  9, 19, 25, 33, 41, 49, 53, 56, 63, 66-8, 78, 106, 125-6, 160, 164-5
RFI (request for inspection)  14, 41, 67-8, 71, 78-9, 82, 90, 94, 99, 103, 114-15, 127
risks  15-22, 29, 32, 49-51, 53-4, 57, 69, 71, 75, 79, 98, 113, 118, 121-2, 151

## S

safety  13-14, 19, 30-1, 45, 49, 56, 58, 62, 69, 73-4, 83, 92-3, 98-101, 119-20, 174-5
service provider  14
Shop Drawings  14, 29, 37, 41, 67-8, 71-2, 94, 99, 106, 116-17, 160, 174, 179
site  26, 48, 51, 55, 57, 66, 72, 74, 76, 78-9, 83-6, 90, 92, 95, 130-1
specifications  11, 19, 29
SPI (schedule performance index)  14, 73, 79, 91, 110, 126, 150
stakeholders  15-17, 19, 45, 96-7
subcontractors  11, 30, 44, 47, 51, 57, 71-2, 74, 80, 95-6, 99, 114, 119, 135, 147
supervision  10, 19, 21-3, 44, 57, 63, 66-7, 127
surveillance  14, 106, 151
SWMS (safe work method statement)  26-7, 30, 50, 52, 54-5, 67, 71, 79, 91, 94, 98-9, 103, 114, 118-20, 127-8

## T

TAB (testing, adjusting, and balancing)  15, 164, 167-9, 171
terms  11-12, 16, 19-20, 42-3, 47, 53-4, 61-3, 123, 125, 131, 136-7, 147, 150, 152, 185
tests  22, 90, 157, 159-60, 162, 165-7, 169-70, 172
tolerance  15, 29, 36, 38, 46, 62, 75, 78
top 20 per cent  15, 135-7, 144-5, 147
training  22, 27, 57, 70, 72, 79, 84, 99, 102, 110, 115, 120, 122, 146

## V

verification  42, 63, 89, 155, 157, 164, 166, 170-3, 175

## W

warranties  173, 181, 183
WBS (work breakdown structure)  16, 42, 128, 130, 146
witness point  13, 16, 116-17
work packages  16, 42-4, 128, 130, 154
workers  20, 26-7, 49, 84-5, 92, 116
workplace  54, 69, 71, 121

# About the Book and Authors

This is the third book of the series that has documented best practice within the building industry, detailing the many processes required to procure buildings. The first book, titled *City of Layers: Reconfiguring the Built Environment for Sustainability*, outlines how buildings can be procured sustainably. The second book, titled *The Project Manager's Checklist for Building Projects, Delivery Strategies and Processes*, details how design and procurement processes should be planned and managed. This third book focuses on construction, post-design, and procurement. It outlines how the planned and desired outcome in terms of quality can be achieved safely whilst minimising harm to the environment.

Each book was formatted for operational use for specific projects, providing a roadmap of information with checklists that also doubles as a valuable and portable paper trail, adding value to the project's quality assurance processes. On completion of project, this book, complete with project notes, can provide a historical record of what was considered and what was done at each phase of the project life.

This third book, titled *Construction Supervision: QC + HSE Management in Practice*, details the in-practice monitoring and controlling aspects of construction works. It outlines what should be considered as the supervision process is planned and what should be actioned as construction works proceeds. In writing this book, the hope is such knowledge will enable practitioners to focus on doing the required things and ensuring the things are done right so construction liabilities and risks are minimised/mitigated, whilst adding substantial value to what is done, benefiting both the project and society at large.

**Mark Urizar**
FAIA, B.Arch, PMP, MBA, MAppSc, Leed-AP (Australia)
Architect, project and design manager

**El-Sayed Abdel Monem Sayed Abdel Halim**
BSc Civil, PMP, MAPM, PE (Egypt)
Engineer, project manager

www.ingramcontent.com/pod-product-compliance
Lightning Source LLC
Chambersburg PA
CBHW030940180526
45163CB00002B/639